High Speed Flow

High Speed Flow is a textbook suitable for undergraduates, postgraduates, and research workers in fluid dynamics. It covers such topics as subsonic and supersonic flight, shock waves, high-speed aerofoils, and thermodynamics. Starting from first principles, the book gives complete and elementary derivations of all results and takes the reader to research level in the subject.

The book contains many exercises and an extensive bibliography, providing access to the entire literature of the subject from 1860 to the present day, and including over two hundred items published since 1990. It contains the most extensive set of formulae on oblique shock waves ever assembled.

Dr. Chapman is Reader in Mathematics and Mechanics at the University of Keele, England. He studied mathematics at the University of Cambridge and received his Ph.D. in fluid dynamics from the University of Bristol. He has held a Fellowship at the Woods Hole Oceanographic Institution and worked in the nuclear power industry, and he was a Fellow of Christ's College, Cambridge.

Cambridge Texts in Applied Mathematics

High Speed Flow

C. J. CHAPMAN

University of Keele, England

CAMBRIDGE UNIVERSITY PRESS
Cambridge, New York, Melbourne, Madrid, Cape Town, Singapore,
São Paulo, Delhi, Dubai, Tokyo, Mexico City

Cambridge University Press
The Edinburgh Building, Cambridge CB2 8RU, UK

Published in the United States of America by Cambridge University Press, New York

www.cambridge.org
Information on this title: www.cambridge.org/9780521666473

First published 2000

A catalogue record for this publication is available from the British Library

Library of Congress Cataloguing in Publication Data
Chapman, C. J. (Christopher John), 1952–
High speed flow / C.J. Chapman
p.   cm. – (Cambridge texts in applied mathematics)
Includes bibliographical references and index.
ISBN 0-521-66169-2 (hbk.). – ISBN 0-521-66647-3 (pbk.)
1. Fluid dynamics. I. Title. II. Series.
QA911.C43   2000
532'.0532 – dc21   99-37544
                                    CIP

ISBN 978-0-521-66169-0 Hardback
ISBN 978-0-521-66647-3 Paperback

Contents

Contents ix

Preface

This book is based on a course of lectures I gave for several years to students reading Part III of the Mathematical Tripos at the University of Cambridge. The course was well received, and since giving the course I have had requests for copies of the lecture notes. The students on the course were familiar with the basics of fluid dynamics, but they had little prior knowledge of the effects of compressibility and required a course that, though elementary, would prepare them for the literature of high speed flow in the larger texts and in research journals, for example in the *Journal of Fluid Mechanics*.

In writing the book I have kept closely to the topics included in the original lecture course, which consisted of sixteen one-hour lectures. Accordingly, the book is suitable for undergraduate or beginning graduate students and could be studied subject-by-subject or in its entirety. The book derives most results from first principles, so that the only formal prerequisite from fluid dynamics is familiarity with the equations of motion (i.e., the equations of conservation of mass, momentum, and energy). All results from thermodynamics are derived from first principles. The emphasis is on topics that occur as simple components in the description of general flows; these topics include the Rankine–Hugoniot relation, Prandtl–Meyer expansion, Prandtl's relation, Riemann invariants, Mach reflection, von Neumann reflection, Maxwell's relations, Legendre transformations, the Laval nozzle, Hamilton's ray equations, the Monge cone, Crocco's equation, Ringleb's flow, Helmholtz free energy, and Gibbs free energy. The book contains a large number of exercises and problems for the reader, which would also be suitable for a lecturer in setting problem sheets or examinations.

After deliberation, I decided to include in the book a set of references large enough to provide an entry point to the entire literature of high speed flow, but to place all references in special sections at the end of each chapter and in a special chapter at the end of the book. In particular, I have included references to all the papers on high speed flow published in the *Journal of Fluid Mechanics* in the period 1990–1998. These papers, somewhat over two hundred in number, lend themselves readily to classification into four main groups, indicated by the tables in Section 13.6, and give a fair indication of the scope of current research in high speed flow. After studying this elementary book, the reader may be pleasantly surprised to find that an understanding of nearly all these papers is within his or her grasp.

I thank many colleagues, in the United Kingdom and throughout the world, for their generosity in providing comments on individual chapters of the book.

1

Preliminaries

1.1 The Mach Number

The theory of high speed flow is concerned with flows of fluid at speeds high enough that account must be taken of the fluid's compressibility. The theory finds application in many branches of science and technology, from which we may single out, as being of unrivalled importance in the modern world, the applications to high speed flight.

The dimensionless parameter that measures the importance of a fluid's compressibility in high speed flow is the *Mach number.* Suppose that, at a given point in space and time, the speed of the fluid is u and the speed of sound is c. The Mach number M is defined as the ratio u/c. Thus

$$M = \frac{u}{c}. \qquad (1.1.1)$$

In general, the value of the Mach number varies with position and time. But in many problems, we may choose a representative flow speed, say U, and a representative sound speed, say c. Then the quantity U/c is a single number measuring the importance of compressibility in the flow, and we may say that the flow is taking place at a Mach number $M = U/c$.

1.2 Flow Regimes

Flows corresponding to different ranges of the Mach number M have very different properties. We shall distinguish five regimes, namely (a) incompressible flow, (b) subsonic flow, (c) transonic flow, (d) supersonic flow, and (e) hypersonic flow.

(a) Incompressible Flow

This regime is defined by $M \ll 1$. The flow speed is low enough for the fluid's compressibility to be negligible. The regime is the basis of the classical subject

1

of hydrodynamics. Simple though it seems to state that $M \ll 1$, there are nevertheless some subtleties in the limit $M \to 0$. For example, the flow corresponding to the limit (U fixed, $c \to \infty$) can differ from that corresponding to ($U \to 0, c$ fixed), because the former filters out sound waves, whereas the latter does not. Perhaps (U fixed, $c \to \infty$) should be called the incompressible limit, and ($U \to 0, c$ fixed) the low speed limit.

(b) Subsonic Flow

This regime is defined by $0 < M < 1$, subject to the restriction that M is not too close to 0 or 1. The flow speed is high enough for the fluid's compressibility to be important, but low enough for the speed to be comfortably clear of the speed of sound. Since most aircraft fly well below the speed of sound, the regimes of incompressible flow and subsonic flow include most of standard aeronautics. The subsonic regime includes much of acoustics.

(c) Transonic Flow

This regime is defined by M being close to 1 in some important part of the flow. The regime raises difficult and interesting mathematical questions, because the governing partial differential equations are then of mixed type. That is, in some regions the equations are elliptic, in other regions they are hyperbolic, and on the separating lines or surfaces they are parabolic. The transonic regime is of vital importance to an aircraft or a land vehicle that "breaks the sound barrier."

(d) Supersonic Flow

This regime is defined by $M > 1$, subject to the restriction that M is not too close to 1 nor too large. Parts of the mathematical theory of supersonic flow can be obtained from that for subsonic flow by replacing $(1 - M^2)^{\frac{1}{2}}$ and an elliptic equation by $(M^2 - 1)^{\frac{1}{2}}$ and a hyperbolic equation. The theory of the supersonic regime was of importance to the design of the civil supersonic airplane *Concorde*.

(e) Hypersonic Flow

This regime is defined by $M \gg 1$. The flow speed is so high that compressibility is all-important, particularly in producing very high temperatures and ionization. The hypersonic regime is of importance for rockets and for the civil hypersonic aircraft "Orient Express."

In formulae for subsonic flow, two factors that often appear are the Doppler factor $(1 - M^2)^{\frac{1}{2}}$ and a factor that for air takes the value $(1 + \frac{1}{5}M^2)^{\frac{1}{2}}$. In many practical problems, the flow may be regarded as effectively incompressible if the relevant factor is within a few percent of unity. For example, in air at atmospheric pressure and a temperature of 20°C, the speed of sound is $340 \, \mathrm{m \, s}^{-1}$. Thus a speed of 100 m s^{-1} corresponds to $M = 0.294$, for which $(1 - M^2)^{\frac{1}{2}} = 0.956$ and $(1 + \frac{1}{5} M^2)^{\frac{1}{2}} = 1.009$; and a speed of 150 m s^{-1} corresponds to $M = 0.441$, for which $(1 - M^2)^{\frac{1}{2}} = 0.898$ and $(1 + \frac{1}{5} M^2)^{\frac{1}{2}} = 1.019$. Therefore, at sea level, compressibility is important at speeds of 100 m s^{-1}, and in precise work it may be important at somewhat lower speeds. For an aircraft flying at a given speed, the Mach number increases with height, because in the lower atmosphere the speed of sound decreases with height.

A feature of high speed flow in all except the incompressible and subsonic regimes is the widespread occurrence of shock waves, that is, surfaces across which the fluid may be regarded as having a discontinuity in pressure. The theory of shock waves is an important part of the subject of high speed flow and occupies an appreciable proportion of this book.

1.3 Temperature Changes

In high speed flow, the variables in the momentum equation are coupled to the thermodynamic variables, because changes in pressure compress or expand the fluid and alter its temperature. Equally, changes in temperature affect the pressure, via the equation of state. Therefore in the study of high speed flow there is no escape from some thermodynamics. In particular, it is necessary at some stage to introduce the concept of entropy.

In this book, the required thermodynamic theory is elementary and is derived from first principles. The main facts used are the definitions and basic properties of specific heats, enthalpy, and the gas constant. Some aspects of high speed flow, for example the theories of hypersonic flow and flow with combustion, require advanced ideas from thermodynamics, and thus they are beyond the scope of this book.

1.4 History

The most striking work on high speed flow in the nineteenth century was the visualisation of shock waves in air. In the 1870s, Mach deduced their positions from the white lines where intersections of shock waves blow the soot off a sooted glass plate, and hence he discovered many of their properties, including the different types of reflection, V-shaped or Y-shaped, that are possible at a wall. The shock waves were generated by electric sparks. In the 1880s, Mach and Salcher photographed the shock wave ahead of a supersonic projectile. They used the schlieren method, invented in 1864 by Toepler, in which changes in refractive index due to changes in density are made visible by a special type of illumination. The photographs were excellent and showed, as well as the shock wave, many details of the flow, such as the "Mach lines" produced by surface roughness and the turbulent wake. Mach and Salcher also observed the diamond-shaped pattern of shock waves that can occur with a supersonic jet. The ratio u/c had been introduced to high speed flow by Doppler in the 1840s, before Mach's scientific work, and was not called the Mach number during Mach's lifetime (1838–1916); it was first called the Mach number by Ackeret in 1929. Mach was primarily an experimentalist and a philosopher, and some of his ideas must sound strange to most fluid dynamicists: For example, he did not believe in atoms and molecules. But Einstein was deeply influenced by Mach and publicized "Mach's principle."

In the twentieth century, the years prior to the Second World War saw steady progress in the theory and practice of high speed flow, especially by German engineers. High speed flow became a particularly important subject during 1939–1945 not only because of the war but also because at that time aircraft speeds were approaching the speed of sound. An intense research effort into the technology of high speed flight and ballistics was stimulated and drew upon the efforts of some powerful mathematicians, including Lighthill and von Neumann. The theory of characteristics reached its modern form during the war years and shortly afterwards. Other mathematical techniques, based on perturbation theory, were steadily extended as they were applied to problems of greater difficulty, and their modern form emerged only some years later with the development of, for example, the theory of matched asymptotic expansions.

A feature of research work on high speed flow since the Second World War has been the increasing use of high speed computers, hand-in-hand with the creation of a new subject, computational fluid dynamics. Among many successes has been the numerical computation of transonic flow fields, as required for the design of transonic aerofoils. The use of high speed computers now pervades all aspects of research into high speed flow and, indeed, other types of flow.

1.5 Recent Research

A large amount of research activity is currently taking place worldwide on problems related to high speed flow. The research takes place in university departments of mathematics, engineering, physics, and chemistry; in government research centers, for example in England at the Defence and Evaluation Research Agency, Farnborough, and in the United States at NASA Ames, Langley, and Lewis; and in corporate research centers, for example in England at Rolls-Royce, Derby and in the United States at research centers connected with Boeing, General Electric, McDonnell Douglas, Pratt and Whitney, United Technologies, and their various merged entities. An inspection of articles in research journals, for example the *Journal of Fluid Mechanics*, shows that, in the several decades up to 1998, particularly active research areas related to high speed flow include (a) shock waves, (b) hypersonic flow, (c) jets, boundary layers, shear layers, and mixing layers, (d) high speed propellers and turbines, (e) aeroacoustics, and (f) combustion. The practical and commercial importance of research in these areas is confirmed by numerous articles in the aerospace magazine *Aviation Week*. Much research work is concerned with combinations of topics taken from areas (a)–(f), for example shock wave/boundary layer interactions, or the aeroacoustics of jets, or combustion in hypersonic flow. The number of such combinations having practical importance is large; and each combination can be related to the three classical fluid dynamical subjects of waves, stability, and turbulence and to the three classical methods of experiment, theory, and numerical computation. In this book we shall not try to survey such an enormous area of work. But throughout the book we give references to a representative sample of recent research papers in areas (a)–(f). We shall resist the temptation to extend the list of research areas even further, for example to magnetohydrodynamics and astrophysics.

1.6 Bibliographic Notes

Some articles on the history and practioners of high speed flow are: "Contributions of Ernst Mach to fluid dynamics," Reichenbach (1983); "Jakob Ackeret and the history of the Mach number," Rott (1985); "Compressible flow in the thirties," Busemann (1971); "Recalling the Vth Volta congress: High speeds in aviation," Ferrari (1996); and "Keith Stewartson: His life and work," Stuart (1986). The study of shock waves by the white lines they produce on a sooted plate is described in Mach (1878), and photographs of shock waves appear in Mach & Salcher (1887). A list of selected early papers relating to high speed flow, from Earnshaw (1860) to Frankl (1945), is given in Table 13.2.1 of Chapter 13. The table lists the founders of the subject, and a perusal of the titles of the papers, in the references, will give a preliminary idea of the way the subject developed. Other contributors to research in high speed flow, now better known for their work in other fields, were Chandrasekhar, Gamow, Robinson, Shepherdson, Tukey, and Weyl.

Much of the research work on high speed flow performed in the war years 1939–1945 and shortly afterwards appeared in systematic form in the monograph "Supersonic flow and shock waves," Courant & Friedrichs (1948) and in two reference works, all with extensive bibliographies. The first reference work was published in 1953 as the two volumes of *Modern Developments in Fluid Dynamics: High Speed Flow*, edited by Howarth. The second was published during the period 1955–1964 as the twelve volumes of *High Speed Aerodynamics and Jet Propulsion*. Each volume had its own editors, and the series as a whole was edited at different times by Summerfield, Charyk, and Donaldson.

Another reference work is the *Encyclopedia of Physics*, edited by Flügge. The volumes containing articles on high speed flow were published in 1959 as *Fluid Dynamics I* and in 1960 as *Fluid Dynamics III*, forming Volume 8 Part I and Volume 9 of the encyclopedia. The volumes were coedited by Truesdell.

The three reference works are each the work of many authors, and together with Courant & Friedrichs (1948) they present the theory of high speed flow in the form used by later writers and researchers. The works still appear modern. A selection of articles, from Bickley (1953) to Moore (1964), is given in Table 13.3.1. The individual volumes are listed on the first page of the references. Their contents indicate the great increase in knowledge of high speed flow that had been obtained in the period 1939–1964.

The requirements of teaching and research in high speed flow, after the war, led to the production of textbooks and monographs, particularly in the 1950s. A selection of these, from Sauer (1947) to Laney (1998), is given in Table 13.4.1.

Many surveys and reviews of research on high speed flow performed since the mid-1960s appear in the *Annual Review of Fluid Mechanics*. The first volume, published in 1969, contained "Shock waves and radiation" by Zel'dovich & Raizer, and the 1999 volume contains "Computational fluid dynamics of whole-body aircraft" by Agarwal. A selection of these reviews, together with some from other sources, is listed in Table 13.5.1. The first article, a survey of previous work on high speed flow, is by Lighthill (1949), and the next two items, on transonic flow, are by Germain & Bader (1952) and Guderley (1953). Fifteen items are listed for the period 1993–1999.

The papers on high speed flow that have appeared in the *Journal of Fluid Mechanics* in the period 1990–1998 are listed in the tables in Section 13.6. Many papers on high speed flow may be found in Collected Papers of Sir James Lighthill (1997), edited by Hussaini, and many excellent photographs are in *An Album of Fluid Motion*, assembled by Van Dyke (1982).

2

Governing Equations

2.1 Conservation of Mass. Jump Condition

In this chapter we apply conservation laws to arbitrary volumes of fluid, to obtain integral forms of the equations of motion. We consider carefully the conditions under which the integrals can be differentiated to give differential equations of motion, and we show that, at a surface of nonsmoothness, where differentiation is not possible, the integrals nevertheless determine jump conditions relating conditions on opposite sides of the surface. We then discuss the most useful forms of the equations of motion as required in problems of high speed flow.

Let a volume fixed in space be denoted V, and let its fixed surface be S. The volume contains fluid that at position $\mathbf{x}$ and time t has density $\rho(\mathbf{x}, t)$ and velocity $\mathbf{u}(\mathbf{x}, t)$. The equation of conservation of mass, in integral form, asserts that the rate of change of the mass of fluid in V equals the rate at which mass flows into V through S. Let an element of surface area be dS, always taken to be positive, and let the unit normal on S pointing out of V be $\mathbf{n}$, so that the vector element $d\mathbf{S}$ of surface area is $d\mathbf{S} = \mathbf{n}\, dS$, also pointing out of V, and the volume of fluid per unit time leaving V through dS is $\mathbf{u} \cdot d\mathbf{S}$. The definitions are illustrated in Figure 2.1.1. The mass of fluid per unit time leaving V through dS is $\rho\mathbf{u} \cdot d\mathbf{S}$, and the mass of fluid in an element of volume dV is $\rho\, dV$. Thus the equation of conservation of mass is

$$\frac{d}{dt} \int_V \rho\, dV = -\int_S \rho\, \mathbf{u} \cdot d\mathbf{S}. \qquad (2.1.1)$$

The generalization of (2.1.1) to allow for mass sources, as required in the theory of aeroacoustics or combustion for example, is readily obtained by adding to the right-hand side a source integral over V but will not be needed in this book.

If ρ and $\mathbf{u}$ are continuously differentiable functions of $\mathbf{x}$ and t, then in the surface integral on the right of (2.1.1) we may apply the divergence theorem,

7

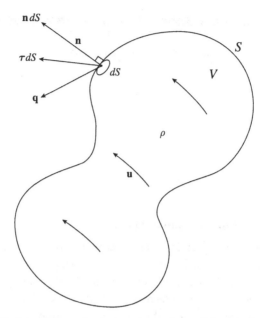

Fig. 2.1.1. Fluid of density ρ and velocity $\mathbf{u}$ in a volume V, fixed in space, bounded by a fixed surface S. At the element of surface area dS, the outward normal is $\mathbf{n}$, the stress is τ, and the heat flux is $\mathbf{q}$. The vector element of area is $\mathbf{n}\,dS$, and the force at dS on the fluid in V is $\tau\,dS$.

and in the volume integral on the left we may take the differentiation with respect to t under the integration sign. Thus

$$\int_V \left\{ \frac{\partial \rho}{\partial t} + \nabla \cdot (\rho \mathbf{u}) \right\} dV = 0. \qquad (2.1.2)$$

Since (2.1.2) holds for an arbitrary volume V, the integrand is zero. Therefore

$$\frac{\partial \rho}{\partial t} + \nabla \cdot (\rho \mathbf{u}) = 0. \qquad (2.1.3)$$

Expanded, (2.1.3) is

$$\frac{\partial \rho}{\partial t} + \mathbf{u} \cdot \nabla \rho + \rho \nabla \cdot \mathbf{u} = 0. \qquad (2.1.4)$$

The convective derivative D/Dt is defined by

$$\frac{D}{Dt} \equiv \frac{\partial}{\partial t} + \mathbf{u} \cdot \nabla. \qquad (2.1.5)$$

Thus (2.1.4) is

$$\frac{D\rho}{Dt} + \rho\nabla \cdot \mathbf{u} = 0. \qquad (2.1.6)$$

Equation (2.1.3), and its equivalent forms (2.1.4) and (2.1.6), are differential versions of the equation of conservation of mass.

Now suppose that ρ and $\mathbf{u}$ are no longer continuously differentiable functions of $\mathbf{x}$ and t, but may be discontinuous, for example because of the presence of a shock or a vortex sheet. Then the integral version (2.1.1) of the equation of conservation of mass remains valid, but not the step from (2.1.1) to (2.1.2) at the surface of discontinuity. Therefore the differential versions of the equation of conservation of mass need to be supplemented by a further equation.

Figure 2.1.2 shows a fixed volume V containing a moving surface of discontinuity. The density and velocity of the fluid in the two parts of V are $\rho_1, \mathbf{u}_1$, and $\rho_2, \mathbf{u}_2$. The differential versions of the equation of conservation of mass apply separately to $\rho_1, \mathbf{u}_1$, and to $\rho_2, \mathbf{u}_2$, and the supplementary condition we need must relate $\rho_1, \mathbf{u}_1$ to $\rho_2, \mathbf{u}_2$ on opposite sides of the moving surface. The limiting values of $\rho_1, \mathbf{u}_1$ and $\rho_2, \mathbf{u}_2$ as the surface is approached are denoted $\rho^-, \mathbf{u}^-$ and $\rho^+, \mathbf{u}^+$.

To obtain the required condition, we use an argument based on the construction of a small cylinder. Consider a fixed cylinder of small radius and still smaller depth, so oriented that its end faces are parallel to a surface of discontinuity that may be moving through it. A unit normal to the surface of discontinuity, pointing away from the fluid with density ρ^- and velocity $\mathbf{u}^-$, and towards the fluid with density ρ^+ and velocity $\mathbf{u}^+$, is denoted $\mathbf{n}$. Thus $\mathbf{n}$ is normal to the end faces of the cylinder. The velocity of the moving surface of discontinuity is $\mathbf{V}$. We allow $\mathbf{V}$ to point in any direction, although the final formulae depend on $\mathbf{V}$ only through $\mathbf{n} \cdot \mathbf{V}$. For any quantity associated with the fluid, the jump in value across the moving surface, from the negative side to the positive side, is denoted by square brackets, so that, for example,

$$[\rho] = [\rho]_-^+ = \rho^+ - \rho^-. \qquad (2.1.7)$$

The volume of the cylinder is V, and the area of each end face is A. The definitions are illustrated in Figure 2.1.3.

We now evaluate, for the cylinder, each side of the integral version of the equation of conservation of mass, (2.1.1). Since fluid of density ρ^+ is being replaced with fluid of density ρ^- by a surface moving at a speed normal to itself of $\mathbf{n} \cdot \mathbf{V}$, and $\mathbf{n} \cdot \mathbf{V}$ measures the volume swept out per unit time by unit area of the surface of discontinuity, it follows that the change per unit time of the mass

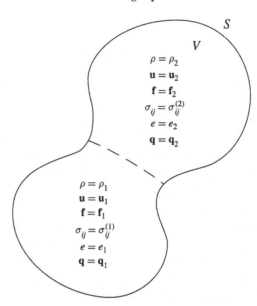

Fig. 2.1.2. Volume V, fixed in space, bounded by a fixed surface S and containing a moving surface of discontinuity ($-\,-$). Fluid density ρ, fluid velocity $\mathbf{u}$, body force per unit mass $\mathbf{f}$, stress tensor σ_{ij}, internal energy per unit mass e, heat flux $\mathbf{q}$; opposite sides of the surface of discontinuity are denoted 1, 2.

of fluid in the cylinder is given by

$$\frac{d}{dt} \int_V \rho \, dV = (\rho^- - \rho^+)(\mathbf{n} \cdot \mathbf{V})A$$
$$= -[\rho](\mathbf{n} \cdot \mathbf{V})A. \tag{2.1.8}$$

The depth of the cylinder is assumed small enough compared with its radius that the mass flow through the curved surface is negligible compared with that through the end faces. Hence

$$\int_S \rho \mathbf{u} \cdot d\mathbf{S} = \{\rho^+(\mathbf{u}^+ \cdot \mathbf{n}) - \rho^-(\mathbf{u}^- \cdot \mathbf{n})\}A$$
$$= [\rho \mathbf{u} \cdot \mathbf{n}]A. \tag{2.1.9}$$

Therefore conservation of mass gives

$$[\rho](\mathbf{n} \cdot \mathbf{V}) = [\rho \mathbf{u} \cdot \mathbf{n}]. \tag{2.1.10}$$

Thus

$$[\rho(\mathbf{u} - \mathbf{V}) \cdot \mathbf{n}] = 0. \tag{2.1.11}$$

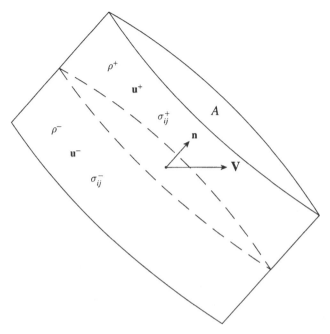

Fig. 2.1.3. Small fixed cylinder of fluid containing a surface of discontinuity moving at velocity **V**. The area of the end face of the cylinder is A; opposite sides of the surface of discontinuity are denoted $-$, $+$; the unit normal to the surface of discontinuity, pointing from the negative side to the positive side, is **n**.

Equation (2.1.11) is the jump condition representing conservation of mass across a surface of discontinuity.

A second method of obtaining (2.1.11) is to consider a small cylinder of constant size moving with the surface of discontinuity. Relative to the cylinder, the surface of discontinuity is at rest, and the normal velocity of the fluid jumps from $(\mathbf{u}^- - \mathbf{V}) \cdot \mathbf{n}$ to $(\mathbf{u}^+ - \mathbf{V}) \cdot \mathbf{n}$. Conservation of mass gives (2.1.11) at once. The second method is quicker than the first, but it is harder to apply generally, for example to conservation of momentum, because surfaces of discontinuity usually accelerate. Therefore the frame of reference moving with a cylinder attached to a surface of discontinuity is not an inertial frame. The second method thus requires consideration of "fictitious forces," even if these do not affect the final answer. The first method applies generally, because we always have directly an integral form of a conservation law in an inertial frame.

A third method of obtaining jump conditions is to use volumes that move with the fluid, so that they always contain the same fluid particles. Such volumes automatically contain a constant mass of fluid. We illustrate the method by

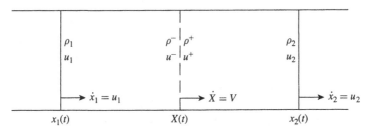

Fig. 2.1.4. One-dimensional flow. The region $x_1(t) \leq x \leq x_2(t)$ moves with the fluid and has a surface of discontinuity at $x = X(t)$.

deriving (2.1.11) for one-dimensional flow. Let the position coordinate be x, and let the boundaries of a region moving with the fluid be $x_1(t)$ and $x_2(t)$, so that the moving region is $x_1(t) \leq x \leq x_2(t)$. The fluid has density $\rho(x, t)$ and its velocity in the x direction is $u(x, t)$. The moving surface of discontinuity is at $x = X(t)$, so that $x_1(t) < X(t) < x_2(t)$, and across $x = X(t)$ are discontinuities in ρ and u. The flow is illustrated in Figure 2.1.4.

Since the region $x_1(t) \leq x \leq x_2(t)$ always consists of the same fluid particles, the integral form of the equation of conservation of mass is

$$\frac{d}{dt} \int_{x_1(t)}^{x_2(t)} \rho(x, t) \, dx = 0. \tag{2.1.12}$$

To differentiate the integral we must split up the range (x_1, x_2) into (x_1, X) and (X, x_2) and allow for the dependence on t of x_1, x_2, and X. Thus

$$\frac{d}{dt} \left\{ \int_{x_1}^{X} \rho \, dx + \int_{X}^{x_2} \rho \, dx \right\} = 0. \tag{2.1.13}$$

The values of ρ and u at x_1 and x_2 are ρ_1, ρ_2 and u_1, u_2 respectively, so that

$$
\begin{aligned}
\rho_1 &= \rho_1(t) = \rho(x_1(t), t), \\
\rho_2 &= \rho_2(t) = \rho(x_2(t), t), \\
u_1 &= u_1(t) = u(x_1(t), t), \\
u_2 &= u_2(t) = u(x_2(t), t).
\end{aligned}
\tag{2.1.14}
$$

The limiting values of ρ on the left and right sides of the moving surface $x = X(t)$ are ρ^- and ρ^+, and the jump is $[\rho] = \rho^+ - \rho^-$. Differentiation with respect to t is denoted by a dot, and the velocity of the surface of discontinuity in the x direction is V, so that $V = V(t) = \dot{X}(t)$. On differentiating the integrals in (2.1.13) with respect to t, and using the formula for differentiation of an integral with variable end points, we obtain

$$\frac{d}{dt}\int_{x_1}^{X} \rho \, dx = \int_{x_1}^{X} \frac{\partial \rho}{\partial t} \, dx + \rho^{-}\dot{X} - \rho_1 \dot{x}_1,$$

$$\frac{d}{dt}\int_{X}^{x_2} \rho \, dx = \int_{X}^{x_2} \frac{\partial \rho}{\partial t} \, dx + \rho_2 \dot{x}_2 - \rho^{+}\dot{X}.$$

(2.1.15)

Since the points x_1 and x_2 move with the fluid, we have $\dot{x}_1 = u(x_1(t), t) \equiv u_1(t) \equiv u_1$ and $\dot{x}_2 = u(x_2(t), t) \equiv u_2(t) \equiv u_2$. Substitution of (2.1.15) into (2.1.13) gives

$$\int_{x_1}^{x_2} \frac{\partial \rho}{\partial t} dx - [\rho]V + \rho_2 u_2 - \rho_1 u_1 = 0.$$

(2.1.16)

Here the integral of $\partial \rho / \partial t$ is defined to be the sum of the two separate integrals on the intervals (x_1, X) and (X, x_2), so that $\partial \rho / \partial t$ is bounded, and the point $x = X(t)$ contributes to $\partial \rho / \partial t$ a possible discontinuity but no singularity or delta-function. Therefore on taking the limit as x_1 tends upwards to X, and x_2 tends downwards to X, the integral of $\partial \rho / \partial t$ in (2.1.16) tends to zero. In addition, $(\rho_1, \rho_2) \to (\rho^{-}, \rho^{+})$ and $(u_1, u_2) \to (u^{-}, u^{+})$. Hence the limiting form of (2.1.16) is

$$-[\rho]V + \rho^{+}u^{+} - \rho^{-}u^{-} = 0.$$

(2.1.17)

That is,

$$-[\rho]V + [\rho u] = 0.$$

(2.1.18)

Thus

$$[\rho(u - V)] = 0.$$

(2.1.19)

Equation (2.1.19) is the one-dimensional version of (2.1.11).

The derivation of conservation equations and jump conditions from volumes that move with the fluid applies in any number of dimensions. For example, the time derivative of the single integral in (2.1.12) becomes, in three dimensions, the time derivative of a volume integral. Then in (2.1.15) the terms $-\rho_1 \dot{x}_1$ and $\rho_2 \dot{x}_2$ become a surface integral over the moving boundary of the volume, and $\rho^{-}\dot{X}$ and $-\rho^{+}\dot{X}$ become surface integrals over the two sides of the moving surface of discontinuity inside the volume.

The three methods just described may be used to obtain the jump conditions corresponding to any conserved quantity. The methods are based on cylinders of constant size, fixed in space; of constant size, moving with the surface of discontinuity; and of varying size, containing always the same fluid particles. Depending on the conservation law being considered, one method may be slightly easier to apply than another, but the three methods always give the same answer.

2.2 Conservation of Momentum. Jump Condition

As in Section 2.1, we consider a volume V fixed in space, bounded by a fixed surface S. At position $\mathbf{x}$ and time t, the fluid in V has density $\rho(\mathbf{x}, t)$ and velocity $\mathbf{u}(\mathbf{x}, t)$, and we assume that the fluid is acted on by a body force $\mathbf{f}(\mathbf{x}, t)$ per unit mass, that is, $\rho(\mathbf{x}, t)\,\mathbf{f}(\mathbf{x}, t)$ per unit volume. The equation of conservation of momentum, in integral form, asserts that the rate of change of the momentum of the fluid in V, plus the rate at which momentum leaves V through S, equals the total body force exerted on the fluid in V, plus the total force exerted on the fluid at S.

The force per unit area acting over S on the fluid in V is the stress $\boldsymbol{\tau}(\mathbf{x}, t)$, so that the force acting on the element of area dS is $\boldsymbol{\tau}(\mathbf{x}, t)\,dS$. The unit normal on S pointing out of V is $\mathbf{n}$, the vector element of surface area is $d\mathbf{S} = \mathbf{n}\,dS$, and the volume of fluid per unit time leaving V through dS is $\mathbf{u} \cdot d\mathbf{S}$. Since the momentum of fluid per unit volume is $\rho(\mathbf{x}, t)\,\mathbf{u}(\mathbf{x}, t)$, the momentum per unit time leaving V through dS is $\rho\mathbf{u}(\mathbf{u} \cdot d\mathbf{S})$. The definitions are illustrated in Figure 2.1.1. Thus the equation of conservation of momentum is

$$\frac{d}{dt} \int_V \rho\mathbf{u}\,dV + \int_S \rho\mathbf{u}(\mathbf{u} \cdot d\mathbf{S}) = \int_V \rho\mathbf{f}\,dV + \int_S \boldsymbol{\tau}\,dS. \qquad (2.2.1)$$

The dependence of $\boldsymbol{\tau}$ on $\mathbf{n}$ is expressed by means of the stress tensor $\boldsymbol{\sigma}$. Let coordinates be labeled by suffices 1,2,3, so that $\mathbf{n} = (n_1, n_2, n_3)$, $\boldsymbol{\tau} = (\tau_1, \tau_2, \tau_3)$, and the nine components of $\boldsymbol{\sigma}$ are $(\sigma_{11}, \sigma_{12}, \ldots, \sigma_{33})$. The summation convention will be used for repeated suffices, and the "dot product" of a tensor and a vector will be defined as for the product of a matrix and a vector, by summation over neighbouring suffices. By definition of $\boldsymbol{\sigma}$, the relation between $\boldsymbol{\tau}$ and $\mathbf{n}$ is $\tau_i = \sigma_{ij}n_j$, that is, $\boldsymbol{\tau} = \boldsymbol{\sigma} \cdot \mathbf{n}$. Therefore the last term of (2.2.1) contains $\tau_i dS = \sigma_{ij}n_j dS = \sigma_{ij}dS_j$ in suffix form, or $\tau dS = \boldsymbol{\sigma} \cdot \mathbf{n}\,dS = \boldsymbol{\sigma} \cdot d\mathbf{S}$ in vector form.

If ρ, $\mathbf{u}$, and $\boldsymbol{\sigma}$ are continuously differentiable functions of $\mathbf{x}$ and t, then in (2.2.1) we may apply the divergence theorem and take the differentiation with respect to t under the integration sign. The result is

$$\int_V \frac{\partial}{\partial t}(\rho u_i)\,dV + \int_V \frac{\partial}{\partial x_j}(\rho u_i u_j)\,dV$$
$$= \int_V \rho f_i\,dV + \int_V \frac{\partial}{\partial x_j}(\sigma_{ij})\,dV. \qquad (2.2.2)$$

Since (2.2.2) holds for an arbitrary volume V, we obtain

$$\frac{\partial}{\partial t}(\rho u_i) + \frac{\partial}{\partial x_j}(\rho u_i u_j) = \frac{\partial \sigma_{ij}}{\partial x_j} + \rho f_i. \qquad (2.2.3)$$

Equation (2.2.3), and equivalent forms obtained by subtracting a multiple of the equation of conservation of mass, are differential versions of the equation of conservation of momentum.

Now suppose, as in Section 2.1, that ρ, $\mathbf{u}$, and σ are no longer continuously differentiable functions of $\mathbf{x}$ and t, but may be discontinuous, for example because of the presence of a shock wave or a vortex sheet. Discontinuities in σ arise from discontinuities in τ through the relation $\tau_i = \sigma_{ij} n_j$. The integral relation (2.2.1) remains valid, but not the step to (2.2.2) at the surface of discontinuity. Therefore we need to supplement (2.2.3).

Consider again Figure 2.1.2, showing a fixed volume V containing a moving surface of discontinuity that separates V into parts 1 and 2. Equation (2.2.3) applies to $\rho_1, \mathbf{u}_1, \mathbf{f}_1, \sigma_{ij}^{(1)}$ and to $\rho_2, \mathbf{u}_2, \mathbf{f}_2, \sigma_{ij}^{(2)}$, and we must relate the limiting values $\rho^-, \mathbf{u}^-, \mathbf{f}^-, \sigma_{ij}^-$ and $\rho^+, \mathbf{u}^+, \mathbf{f}^+, \sigma_{ij}^+$ on opposite sides of the moving surface. Take V to be the small fixed cylinder in Figure 2.1.3, and evaluate the terms in (2.2.1). Fluid of momentum per unit volume $\rho^+ \mathbf{u}^+$ is being replaced with fluid of momentum per unit volume $\rho^- \mathbf{u}^-$ by a surface moving with a speed normal to itself of $\mathbf{n} \cdot \mathbf{V}$. Thus

$$\begin{aligned}
\frac{d}{dt}\int_V \rho \mathbf{u}\, dV &= (\rho^- \mathbf{u}^- - \rho^+ \mathbf{u}^+)(\mathbf{n} \cdot \mathbf{V})A \\
&= -[\rho \mathbf{u}](\mathbf{n} \cdot \mathbf{V})A.
\end{aligned} \qquad (2.2.4)$$

Because the depth of the cylinder is much less than its radius, the second term in (2.2.1) is

$$\begin{aligned}
\int_S \rho \mathbf{u}(\mathbf{u} \cdot d\mathbf{S}) &= \{\rho^+ \mathbf{u}^+ (\mathbf{u}^+ \cdot \mathbf{n}) - \rho^- \mathbf{u}^- (\mathbf{u}^- \cdot \mathbf{n})\}A \\
&= [\rho \mathbf{u}(\mathbf{u} \cdot \mathbf{n})]A.
\end{aligned} \qquad (2.2.5)$$

The body force $\mathbf{f}$ per unit mass is assumed bounded. Therefore $\int_V \rho \mathbf{f}\, dV$ is smaller than the other terms in (2.2.1) by a factor proportional to the depth of the cylinder and so is negligible. The i component of $\int_S \tau\, dS$ is

$$\int_S \tau_i\, dS = \int_S \sigma_{ij} n_j\, dS = [\sigma_{ij}]n_j A. \qquad (2.2.6)$$

Therefore (2.2.1) gives

$$-[\rho \mathbf{u}](\mathbf{n} \cdot \mathbf{V}) + [\rho \mathbf{u}(\mathbf{u} \cdot \mathbf{n})] = [\sigma \cdot \mathbf{n}]. \qquad (2.2.7)$$

Thus

$$[\rho\mathbf{u}((\mathbf{u} - \mathbf{V}) \cdot \mathbf{n})] = [\boldsymbol{\sigma} \cdot \mathbf{n}]. \tag{2.2.8}$$

Since (2.2.8) is a relation between vectors, it is equivalent to three independent equations. It is the jump condition representing conservation of momentum across a surface of discontinuity.

The jump condition (2.1.11) for conservation of mass states that $\rho(\mathbf{u} - \mathbf{V}) \cdot \mathbf{n}$, that is, the flux of mass per unit area per unit time, across a surface of discontinuity, is continuous across that surface. Therefore in (2.2.8) we may take the factor $\rho(\mathbf{u} - \mathbf{V}) \cdot \mathbf{n}$ outside of the square brackets, to obtain $(\rho(\mathbf{u} - \mathbf{V}) \cdot \mathbf{n})[\mathbf{u}] = [\boldsymbol{\sigma} \cdot \mathbf{n}]$. Thus (2.2.8) states how the sudden change $[\boldsymbol{\sigma} \cdot \mathbf{n}]$ in the force on the fluid as it passes through the surface of discontinuity produces a sudden change in the velocity of the fluid.

Continuity of $\rho(\mathbf{u} - \mathbf{V}) \cdot \mathbf{n}$ implies continuity of $\rho\mathbf{V}((\mathbf{u} - \mathbf{V}) \cdot \mathbf{n})$, so that $[\rho\mathbf{V}((\mathbf{u} - \mathbf{V}) \cdot \mathbf{n})] = 0$. Subtraction of this equation from (2.2.8) gives

$$[\rho(\mathbf{u} - \mathbf{V})((\mathbf{u} - \mathbf{V}) \cdot \mathbf{n})] = [\boldsymbol{\sigma} \cdot \mathbf{n}]. \tag{2.2.9}$$

Equation (2.2.9) is the form of the jump condition obtained by the second method in Section 2.1. To an observer moving with the surface of discontinuity, the velocity of the fluid is $\mathbf{u} - \mathbf{V}$. Thus (2.2.9) is a transformed version of the statement that, for a surface of discontinuity at rest, the jump condition for conservation of momentum is $[\rho\mathbf{u}(\mathbf{u} \cdot \mathbf{n})] = [\boldsymbol{\sigma} \cdot \mathbf{n}]$.

The third method in Section 2.1 for obtaining a jump condition started with a volume moving with the fluid. For conservation of mass, the absence of mass flow across the boundary of such a volume led in one space dimension to $-[\rho]V + [\rho u] = 0$, that is, $[\rho(u - V)] = 0$. For conservation of momentum, we use the fact that there is no momentum flow $\rho\mathbf{u}(\mathbf{u} \cdot \mathbf{n})$ across such a boundary, although of course the boundary is still acted on by a force, namely the stress $\tau_i = \sigma_{ij}n_j$. In one space dimension, this leads to the jump condition for conservation of momentum in the form

$$-[\rho u]V + [\rho u^2] = [\sigma_{11}], \tag{2.2.10}$$

or

$$[\rho u(u - V)] = [\sigma_{11}]. \tag{2.2.11}$$

2.3 Conservation of Energy. Jump Condition

Consider again a volume V fixed in space, bounded by a fixed surface S, and assume that the fluid in V has internal energy $e(\mathbf{x}, t)$ per unit mass, that is,

$\rho(\mathbf{x}, t) e(\mathbf{x}, t)$ per unit volume. The equation of conservation of energy, in integral form, asserts that the rate of change of energy in V, plus the rate at which energy leaves V through S, equals the rate at which work is done by the body force in V, plus the rate at which work is done by the forces over S, plus the rate of flow of heat into V through S.

The energy per unit volume of fluid, comprising internal energy per unit volume ρe and kinetic energy per unit volume $\frac{1}{2}\rho\mathbf{u}^2$, is $\rho e + \frac{1}{2}\rho\mathbf{u}^2$. Since the volume of fluid per unit time leaving V through dS is $\mathbf{u} \cdot d\mathbf{S}$, it follows that the energy per unit time leaving V through dS is $(\rho e + \frac{1}{2}\rho\mathbf{u}^2)\mathbf{u} \cdot d\mathbf{S}$. The body force per unit volume is $\rho\mathbf{f}$, and the surface force per unit area at S on V is $\boldsymbol{\tau}$. Thus the body force performs work at a rate $\rho\mathbf{u} \cdot \mathbf{f}$ per unit volume, and the surface force performs work at a rate $\mathbf{u} \cdot \boldsymbol{\tau}$ per unit area. The heat flow per unit area per unit time (i.e., the heat flux) is $\mathbf{q}(\mathbf{x}, t)$, so that the heat per unit time entering V through dS is $-\mathbf{q} \cdot d\mathbf{S}$. These definitions are illustrated in Figure 2.1.1. Thus the equation of conservation of energy is

$$\frac{d}{dt} \int_V \left(\rho e + \frac{1}{2}\rho\mathbf{u}^2\right) dV + \int_S \left(\rho e + \frac{1}{2}\rho\mathbf{u}^2\right) \mathbf{u} \cdot d\mathbf{S}$$
$$= \int_V \rho\mathbf{u} \cdot \mathbf{f}\, dV + \int_S \mathbf{u} \cdot \boldsymbol{\tau}\, dS - \int_S \mathbf{q} \cdot d\mathbf{S}. \qquad (2.3.1)$$

The expression in the second term on the right is $u_i \tau_i\, dS = u_i \sigma_{ij} n_j\, dS = u_i \sigma_{ij}\, dS_j$ in suffix form, or $\mathbf{u} \cdot \boldsymbol{\sigma} \cdot d\mathbf{S}$ in tensor form.

If ρ, $\mathbf{u}$, $\boldsymbol{\sigma}$, e, and $\mathbf{q}$ are continuously differentiable functions of $\mathbf{x}$ and t, then (2.3.1) may be written

$$\int_V \frac{\partial}{\partial t}\left(\rho e + \frac{1}{2}\rho\mathbf{u}^2\right) dV + \int_V \nabla \cdot \left(\left(\rho e + \frac{1}{2}\rho\mathbf{u}^2\right)\mathbf{u}\right) dV$$
$$= \int_V \rho\mathbf{u} \cdot \mathbf{f}\, dV + \int_V \nabla \cdot (\mathbf{u} \cdot \boldsymbol{\sigma})\, dV - \int_V \nabla \cdot \mathbf{q}\, dV. \qquad (2.3.2)$$

The symbol $\nabla \cdot (\mathbf{u} \cdot \boldsymbol{\sigma})$ represents the scalar $\partial(u_i \sigma_{ij})/\partial x_j$ and would be written in a form such as $(\mathbf{u} \cdot \boldsymbol{\sigma}) \cdot \overset{\leftarrow}{\nabla}$ if we adhered strictly to our convention about neighboring suffices. Since (2.3.2) holds for an arbitrary volume V, we obtain

$$\frac{\partial}{\partial t}\left(\rho e + \frac{1}{2}\rho\mathbf{u}^2\right) + \nabla \cdot \left(\left(\rho e + \frac{1}{2}\rho\mathbf{u}^2\right)\mathbf{u}\right)$$
$$= \rho\mathbf{u} \cdot \mathbf{f} + \nabla \cdot (\mathbf{u} \cdot \boldsymbol{\sigma}) - \nabla \cdot \mathbf{q}. \qquad (2.3.3)$$

This equation, and equivalent forms obtained by subtracting multiples of the equations of conservation of mass and momentum, are differential versions of the equation of conservation of energy.

Now suppose that ρ, $\mathbf{u}$, σ, e, and $\mathbf{q}$ are no longer continuously differentiable functions of $\mathbf{x}$ and t but may be discontinuous. The integral relation (2.3.1) remains valid, but not the step to (2.3.2) at a surface of discontinuity. Therefore we need to supplement (2.3.3). The first method of Section 2.1, applied to the small fixed cylinder in Figure 2.1.3, gives

$$- \left[\rho e + \tfrac{1}{2}\rho \mathbf{u}^2\right](\mathbf{n} \cdot \mathbf{V}) + \left[\left(\rho e + \tfrac{1}{2}\rho \mathbf{u}^2\right)(\mathbf{u} \cdot \mathbf{n})\right]$$
$$= [\mathbf{u} \cdot \sigma \cdot \mathbf{n}] - [\mathbf{q} \cdot \mathbf{n}]. \qquad (2.3.4)$$

Thus

$$\left[\left(\rho e + \tfrac{1}{2}\rho \mathbf{u}^2\right)(\mathbf{u} - \mathbf{V}) \cdot \mathbf{n}\right] = [(\mathbf{u} \cdot \sigma - \mathbf{q}) \cdot \mathbf{n}]. \qquad (2.3.5)$$

This is the jump condition representing conservation of energy across a surface of discontinuity. Equivalent forms may be obtained by subtracting arbitrary multiples of the jump conditions for conservation of mass and momentum, and the form of the jump condition obtained by the second method in Section 2.1 corresponds to one choice of these multiples. The third method in Section 2.1 gives, in one space dimension, with $\mathbf{u} = (u, 0, 0)$, $\mathbf{V} = (V, 0, 0)$, and $\mathbf{q} = (q, 0, 0)$, the result

$$-\left[\rho e + \tfrac{1}{2}\rho u^2\right]V + \left[\left(\rho e + \tfrac{1}{2}\rho u^2\right)u\right] = [u\sigma_{11} - q], \qquad (2.3.6)$$

that is,

$$\left[\left(\rho e + \tfrac{1}{2}\rho u^2\right)(u - V)\right] = [u\sigma_{11} - q]. \qquad (2.3.7)$$

2.4 Equations in Conservation and Nonconservation Form

The differential equations (2.1.3), (2.2.3), and (2.3.3) have the special property that they determine the corresponding jump conditions by the rule that $\partial/\partial t$ is replaced by $-\mathbf{n} \cdot \mathbf{V}$, and ∇ is replaced by $\mathbf{n}$. The nondifferentiated terms are replaced by zero (i.e., discarded). The rule applies because, in obtaining the differential equations from the original integral relations, the product terms have not been expanded, as they could have been, by the formulae of elementary calculus. Thus the terms in (2.1.3), (2.2.3), and (2.3.3) are written exactly as they appear in the integral relations from which they were obtained. Differential equations written in this way are said to be in *conservation form*.

In practice, the most frequently used differential equations are not in conservation form but instead are (a) the conservation of mass equation (2.1.3); (b) the conservation of momentum equation (2.2.3), less a multiple $\mathbf{u}$ of (2.1.3); and (c) the conservation of energy equation (2.3.3), less the scalar product of

u with (2.2.3), less a multiple $e + \frac{1}{2}\mathbf{u}^2$ of (2.1.3). Equations (a)–(c), with the scalar $\sigma_{ij}\partial u_i/\partial x_j$ denoted $\boldsymbol{\sigma} \cdot \nabla \mathbf{u}$, and the vector $\partial \sigma_{ij}/\partial x_j$ denoted $\nabla \cdot \boldsymbol{\sigma}$, are

$$\frac{\partial \rho}{\partial t} + \mathbf{u} \cdot \nabla \rho + \rho \nabla \cdot \mathbf{u} = 0, \tag{2.4.1a}$$

$$\rho \left(\frac{\partial \mathbf{u}}{\partial t} + \mathbf{u} \cdot \nabla \mathbf{u} \right) = \nabla \cdot \boldsymbol{\sigma} + \rho \mathbf{f}, \tag{2.4.1b}$$

$$\rho \left(\frac{\partial e}{\partial t} + \mathbf{u} \cdot \nabla e \right) = \boldsymbol{\sigma} \cdot \nabla \mathbf{u} - \nabla \cdot \mathbf{q}. \tag{2.4.1c}$$

Information about jumps has been lost in obtaining (2.4.1) from (2.1.3), (2.2.3), (2.3.3), because (2.4.1) is not close enough to the integral relations that apply irrespective of smoothness. The information is restored on supplementing (2.4.1) by the jump conditions.

In numerical work, equations in conservation form are useful because they can form the basis of a finite difference scheme that simultaneously represents the smooth part of a solution and any jumps. This is convenient because the position of shock waves, for example, is not usually known in advance.

In analytical work, it is more often convenient to use equations in nonconservation form. The reason is that, as we shall see later, (2.4.1c) makes explicit the role of dissipative processes, such as viscosity and thermal conductivity, in transferring energy from one form to another and increasing entropy. When dissipative processes are negligible, (2.4.1c) is particularly simple. Therefore in this book we shall most often use (2.4.1). For convenience we shall refer to the individual equations in (2.4.1) as the mass equation, the momentum equation, and the energy equation.

2.5 Pressure, Viscosity, and Thermal Conductivity

The mechanical pressure p_m is defined in terms of the diagonal components $\sigma_{11}, \sigma_{22}, \sigma_{33}$ of the stress tensor $\boldsymbol{\sigma}$ by

$$p_m = -\tfrac{1}{3}\sigma_{ii} = -\tfrac{1}{3}(\sigma_{11} + \sigma_{22} + \sigma_{33}). \tag{2.5.1}$$

Define the symbol δ_{ij} by $\delta_{ij} = 1$ when $i = j$ and $\delta_{ij} = 0$ otherwise. For most fluids, including air and water at ordinary temperatures and pressures, $\sigma_{ij} + p_m \delta_{ij}$ is linearly related to the velocity gradients $\partial u_i/\partial x_j$ by a single scalar, the shear viscosity μ. The relation is

$$\sigma_{ij} = -p_m \delta_{ij} + \mu \left(\frac{\partial u_i}{\partial x_j} + \frac{\partial u_j}{\partial x_i} \right) - \frac{2}{3}\mu(\nabla \cdot \mathbf{u})\delta_{ij}. \tag{2.5.2}$$

For the theory that leads to this functional form we refer the reader to the bibliographic notes in Section 2.8. The value of μ depends on other properties of the fluid, most notably temperature, and so in general varies with position and time. Note that σ_{ij} in (2.5.2) is symmetric (i.e., $\sigma_{ij} = \sigma_{ji}$), whereas the most general σ_{ij} in (2.4.1) need not be symmetric if the fluid is subject to a body torque. A fluid satisfying (2.5.2) is said to be Newtonian.

The pressure that appears in thermodynamic relations is the equilibrium pressure p_e. In general, p_e does not equal p_m, because p_m depends only on the translational energy of the molecules, whereas there are time lags while the translational energy reaches equilibrium with other forms of molecular energy. Often, the difference $p_m - p_e$ is linearly related to the velocity gradients $\partial u_i / \partial x_j$ by a single scalar, the bulk viscosity ζ. The relation is

$$p_m = p_e - \zeta \nabla \cdot \mathbf{u}. \tag{2.5.3}$$

The existence of bulk viscosity, being due to a time lag in the attainment of equilibrium in a molecular process, is a relaxation effect. The value of ζ depends on other properties of the fluid and is of the same order of magnitude as the value of μ. But $|\nabla \cdot \mathbf{u}|$ is usually much smaller than the shear rate, that is, the largest of the terms $\partial u_i / \partial x_j$ multiplying μ in (2.5.2). Therefore bulk viscosity may usually be ignored unless there is special reason to include it, for example in the study of dissipation of acoustic energy. Thus when we say that bulk viscosity is negligible, we are usually saying that the viscous effect of $|\nabla \cdot \mathbf{u}|$ is negligible.

The gradient of the temperature T of the fluid will be assumed to be linearly related to the heat flux $\mathbf{q}$ by a scalar, the thermal conductivity k. The relation is

$$\mathbf{q} = -k\nabla T. \tag{2.5.4}$$

Equations such as (2.5.2)–(2.5.4), which depend on the nature of the fluid and contain variables whose values must in general be determined by experiment, are called constitutive relations. We have here presented the most widely used constitutive relations, containing just μ, ζ, and k. These relations need to be modified or extended for complex fluids and various relaxation effects, and for hypersonic flow. Nevertheless they provide an extremely accurate description of a very wide range of phenomena in high speed flow.

2.6 The Navier–Stokes Equations

Variables that specify properties of a substance and can appear in thermodynamic relations are called thermodynamic variables, or variables of state. Those we have met so far are density ρ, temperature T, equilibrium pressure p_e, internal energy per unit mass e, shear viscosity μ, bulk viscosity ζ, and thermal conductivity k. Others we shall meet later include specific enthalpy h and specific entropy s. For the fluids we consider, two thermodynamic variables may be varied independently, and the others are then determined by equations that depend on the nature of the fluid, called equations of state. Let two independent thermodynamic variables be called the selected variables. Any two selected variables, together with the equations of state, determine the other thermodynamic variables. Some useful selected variables are (ρ, T), (p_e, ρ), (p_e, s), and (h, s).

The equations of fluid dynamics contain also nonthermodynamic variables, that is, velocity $\mathbf{u}$, mechanical pressure p_m, stress σ, heat flux $\mathbf{q}$, and body force per unit mass $\mathbf{f}$. The equations may be reduced to five equations in five unknowns (i.e., any two selected variables and any three independent components of $\mathbf{u}$). Thus in the mass, momentum, and energy equations in (2.4.1) we find, on using equations of state as necessary, that any two selected variables and $\mathbf{u}$ determine p_m by (2.5.3), σ by (2.5.2), and $\mathbf{q}$ by (2.5.4). We assume that $\mathbf{f}$ is given. Equations (2.4.1), subject to the constitutive relations (2.5.2)–(2.5.4) in μ, ζ, and k, are the Navier–Stokes equations. When supplemented by initial conditions, boundary conditions, jump conditions, and equations of state, they provide a satisfactory theoretical basis for the analysis of many phenomena in high speed flow. Typical phenomena that lie beyond their reach are ionization effects in hypersonic flow and boundary-layer effects in rarefied gas flow.

We now define some further variables. At position $\mathbf{x}$ and time t, the volume of fluid per unit mass is the specific volume $v(\mathbf{x}, t)$, a thermodynamic variable. By definition, $v(\mathbf{x}, t) = 1/\rho(\mathbf{x}, t)$. Therefore $\rho v = 1$, $D(\rho v)/Dt = 0$, and $\rho^{-1} D\rho/Dt = -v^{-1} Dv/Dt$. In terms of v and $\mathbf{u}$, the mass equation is

$$\frac{\partial v}{\partial t} + \mathbf{u} \cdot \nabla v - v \nabla \cdot \mathbf{u} = 0. \tag{2.6.1}$$

The velocity gradients $\partial u_i / \partial x_j$ define the rate of strain tensor e_{ij} by

$$e_{ij} = \frac{1}{2}\left(\frac{\partial u_i}{\partial x_j} + \frac{\partial u_j}{\partial x_i} \right). \tag{2.6.2}$$

This tensor is symmetric, and its trace is the divergence of the velocity field (i.e., $e_{ij} = e_{ji}$ and $e_{ii} = \nabla \cdot \mathbf{u}$). Subtraction of $\frac{1}{3} e_{kk} \delta_{ij}$ (i.e., $\frac{1}{3}(\nabla \cdot \mathbf{u})\delta_{ij}$) gives the deviatoric part $\tilde{e}_{ij}$ of e_{ij}, that is,

$$\tilde{e}_{ij} = e_{ij} - \tfrac{1}{3}(\nabla \cdot \mathbf{u})\delta_{ij}. \tag{2.6.3}$$

By definition, the deviatoric part of a tensor has trace zero, that is, $\tilde{e}_{ii} = 0$.

We defined the mechanical pressure p_m by $p_m = -\frac{1}{3}\sigma_{ii}$. Therefore $\sigma_{ij} + p_m\delta_{ij}$ is the deviatoric part of σ_{ij}, say $\tilde{\sigma}_{ij}$. That is,

$$\tilde{\sigma}_{ij} = \sigma_{ij} + p_m\delta_{ij}. \tag{2.6.4}$$

Equation (2.5.2) for a Newtonian fluid is

$$\tilde{\sigma}_{ij} = 2\mu\tilde{e}_{ij}. \tag{2.6.5}$$

Thus

$$\sigma_{ij} = -p_m\delta_{ij} + 2\mu\tilde{e}_{ij}. \tag{2.6.6}$$

With $p_m = p_e - \zeta\nabla\cdot\mathbf{u}$, we obtain

$$\sigma_{ij} = -p_e\delta_{ij} + 2\mu\tilde{e}_{ij} + \zeta(\nabla\cdot\mathbf{u})\delta_{ij}. \tag{2.6.7}$$

In the momentum equation, we wrote the vector with components $\partial\sigma_{ij}/\partial x_j$ as $\nabla\cdot\boldsymbol{\sigma}$, and we shall similarly write the vector with components $\partial(2\mu\tilde{e}_{ij})/\partial x_j$ as $\nabla\cdot(2\mu\tilde{e})$. Then (2.6.7) gives

$$\nabla\cdot\boldsymbol{\sigma} = -\nabla p_e + \nabla\cdot(2\mu\tilde{e}) + \nabla(\zeta\nabla\cdot\mathbf{u}). \tag{2.6.8}$$

Hence the momentum equation (2.4.1b) is

$$\rho\left(\frac{\partial\mathbf{u}}{\partial t} + \mathbf{u}\cdot\nabla\mathbf{u}\right) = -\nabla p_e + \nabla\cdot(2\mu\tilde{e}) + \nabla(\zeta\nabla\cdot\mathbf{u}) + \rho\mathbf{f}. \tag{2.6.9}$$

The curl of $\mathbf{u}$ is the vorticity $\boldsymbol{\omega}$, that is,

$$\boldsymbol{\omega} = \nabla\wedge\mathbf{u}. \tag{2.6.10}$$

It is often convenient to write the momentum equation partly in terms of the vorticity by means of the identity

$$\nabla\left(\tfrac{1}{2}\mathbf{u}^2\right) = \mathbf{u}\cdot\nabla\mathbf{u} + \mathbf{u}\wedge\boldsymbol{\omega}. \tag{2.6.11}$$

Thus (2.6.9) is

$$\frac{\partial\mathbf{u}}{\partial t} - \mathbf{u}\wedge\boldsymbol{\omega} = -\frac{1}{\rho}\nabla p_e - \nabla\left(\frac{1}{2}\mathbf{u}^2\right) + \frac{1}{\rho}\nabla\cdot(2\mu\tilde{e})$$
$$+ \frac{1}{\rho}\nabla(\zeta\nabla\cdot\mathbf{u}) + \mathbf{f}. \tag{2.6.12}$$

We shall now rewrite the energy equation in several different ways and deduce in particular that μ, ζ, and k must be nonnegative. When σ is symmetric, we have

$$\sigma_{ij}\frac{\partial u_i}{\partial x_j} = \sigma_{ij}e_{ij}. \tag{2.6.13}$$

Writing σ_{ij} in terms of p_m, and remembering that $\tilde{\sigma}_{ii} = 0$ and $\tilde{e}_{ii} = 0$, we obtain

$$\sigma_{ij}\frac{\partial u_i}{\partial x_j} = (\tilde{\sigma}_{ij} - p_m\delta_{ij})\left(\tilde{e}_{ij} + \frac{1}{3}(\nabla \cdot \mathbf{u})\delta_{ij}\right)$$
$$= -p_m\nabla \cdot \mathbf{u} + \tilde{\sigma}_{ij}\tilde{e}_{ij}. \tag{2.6.14}$$

For a Newtonian fluid with bulk viscosity ζ, the relations $p_m = p_e - \zeta\nabla \cdot \mathbf{u}$ and $\tilde{\sigma}_{ij} = 2\mu\tilde{e}_{ij}$ give

$$\sigma_{ij}\frac{\partial u_i}{\partial x_j} = -p_e\nabla \cdot \mathbf{u} + 2\mu\tilde{e}_{ij}\tilde{e}_{ij} + \zeta(\nabla \cdot \mathbf{u})^2. \tag{2.6.15}$$

We earlier wrote $\sigma_{ij}\partial u_i/\partial x_j$ as $\sigma \cdot \nabla\mathbf{u}$, and we now write $\tilde{e}_{ij}\tilde{e}_{ij}$ as $\tilde{\mathbf{e}}^2$. Then with $\mathbf{q} = -k\nabla T$, the energy equation (2.4.1c) becomes

$$\rho\left(\frac{\partial e}{\partial t} + \mathbf{u} \cdot \nabla e\right) = -p_e\nabla \cdot \mathbf{u} + 2\mu\tilde{\mathbf{e}}^2 + \zeta(\nabla \cdot \mathbf{u})^2 + \nabla \cdot (k\nabla T). \tag{2.6.16}$$

By Equation (2.6.1) in the specific volume v, we have $\nabla \cdot \mathbf{u} = v^{-1}Dv/Dt = \rho Dv/Dt$, so that (2.6.16) is

$$\rho\left(\frac{De}{Dt} + p_e\frac{Dv}{Dt}\right) = 2\mu\tilde{\mathbf{e}}^2 + \zeta(\nabla \cdot \mathbf{u})^2 + \nabla \cdot (k\nabla T). \tag{2.6.17}$$

Assume henceforth that temperature T is measured on the Kelvin temperature scale (i.e., is absolute temperature). We shall see in Chapter 3 that the specific entropy s satisfies the equation

$$T ds = de + p_e\, dv. \tag{2.6.18}$$

Therefore $T Ds/Dt = De/Dt + p_e Dv/Dt$, and (2.6.17) is

$$\rho T\frac{Ds}{Dt} = 2\mu\tilde{\mathbf{e}}^2 + \zeta(\nabla \cdot \mathbf{u})^2 + \nabla \cdot (k\nabla T). \tag{2.6.19}$$

Divide by T and use the identity

$$\nabla \cdot \left(\frac{k\nabla T}{T}\right) = \frac{\nabla \cdot (k\nabla T)}{T} - k\left|\frac{\nabla T}{T}\right|^2. \tag{2.6.20}$$

Then

$$\rho \frac{Ds}{Dt} = \frac{2\mu}{T}\tilde{\mathbf{e}}^2 + \frac{\zeta}{T}(\nabla \cdot \mathbf{u})^2 + k\left|\frac{\nabla T}{T}\right|^2 + \nabla \cdot \left(\frac{k\nabla T}{T}\right). \qquad (2.6.21)$$

Now integrate (2.6.21) over a moving volume V of fluid, containing always the same fluid particles and thermally insulated from its surroundings, so that at the boundary S, on which the unit normal is $\mathbf{n}$, we have $\mathbf{n} \cdot \mathbf{q} = 0$ (i.e., $k\mathbf{n} \cdot \nabla T = 0$). By the divergence theorem, $\nabla \cdot (kT^{-1}\nabla T)$ integrates to zero, to leave

$$\frac{d}{dt}\int_V \rho s \, dV = \int_V \left(\frac{2\mu}{T}\tilde{\mathbf{e}}^2 + \frac{\zeta}{T}(\nabla \cdot \mathbf{u})^2 + k\left|\frac{\nabla T}{T}\right|^2\right) dV. \qquad (2.6.22)$$

We have used a result in vector calculus that the D/Dt operator on the left of (2.6.21), in the presence of the factor ρ, gives the d/dt operator on the left of (2.6.22) when the volume V of integration moves with the fluid. A basic property of entropy, discussed in Chapter 3, shows that for a thermally insulated region the left-hand side of (2.6.22) is nonnegative. Since $T > 0$, by the third law of thermodynamics, we deduce from the squared terms on the right of (2.6.22) that μ, ζ, and k are always nonnegative.

We shall usually ignore viscous effects. In particular, it follows from (2.5.3) that without bulk viscosity there is no difference between the mechanical pressure p_m and the equilibrium pressure p_e. They will each simply be called "the pressure" and denoted p. That is, henceforth we put

$$p = p_m = p_e. \qquad (2.6.23)$$

When viscous and thermal conduction effects are negligible, we may put μ, ζ, and k equal to zero in the Navier–Stokes equations, to obtain the Euler equations. Then the mass, momentum, and energy equations are

$$\frac{\partial \rho}{\partial t} + \mathbf{u} \cdot \nabla \rho + \rho \nabla \cdot \mathbf{u} = 0, \qquad (2.6.24a)$$

$$\rho\left(\frac{\partial \mathbf{u}}{\partial t} + \mathbf{u} \cdot \nabla \mathbf{u}\right) + \nabla p = \rho \mathbf{f}, \qquad (2.6.24b)$$

$$\frac{\partial s}{\partial t} + \mathbf{u} \cdot \nabla s = 0. \qquad (2.6.24c)$$

In obtaining (2.6.24c) we have used (2.6.19). These are the governing equations we shall use most frequently.

2.7 Three Types of Compressible Flow

Three important types of compressible flow, arising from different dominant balances in the governing equations, are (a) high speed flow, (b) sound waves, and (c) flow affected by scale height.

(a) High Speed Flow

Assume that the momentum equation contains a dominant balance between inertia and pressure gradient, that is, between $\mathbf{u} \cdot \nabla \mathbf{u}$ and $\rho^{-1}\nabla p$. Thus if we determined the orders of magnitude of all the terms in the momentum equation, and retained only the largest, then among the terms retained would be $\mathbf{u} \cdot \nabla \mathbf{u}$ and $\rho^{-1}\nabla p$. Often, these would be the only terms retained.

Consider a flow with typical length scale l_0, velocity component u_0, density ρ_0, density difference $\Delta\rho$, specific volume v_0, difference in specific volume Δv, pressure difference Δp, and sound speed c_0. Assume that a dominant term in $\mathbf{u} \cdot \nabla \mathbf{u}$ is of order u_0^2/l_0. The balance between $\mathbf{u} \cdot \nabla \mathbf{u}$ and $\rho^{-1}\nabla p$ gives

$$\frac{u_0^2}{l_0} \sim \frac{\Delta p}{\rho_0 l_0}. \tag{2.7.1}$$

That is,

$$\Delta p \sim \rho_0 u_0^2. \tag{2.7.2}$$

We shall see later that Δp and $\Delta\rho$ are related by

$$\frac{\Delta p}{\Delta\rho} \sim c_0^2. \tag{2.7.3}$$

Elimination of Δp gives

$$\frac{\Delta\rho}{\rho_0} \sim \frac{u_0^2}{c_0^2}. \tag{2.7.4}$$

A typical Mach number M_0 is

$$M_0 = \frac{u_0}{c_0}. \tag{2.7.5}$$

Therefore the definition $v = 1/\rho$ gives

$$\frac{\Delta v}{v_0} \sim \frac{\Delta\rho}{\rho_0} \sim M_0^2. \tag{2.7.6}$$

The dimensionless quantity $v_0^{-1} \Delta v$ is the proportional change in the volume of a fluid element during the flow (i.e., the compression of the fluid). Thus under the assumptions stated, the compression varies as the square of the Mach number. This important fact implies that in high speed flow the Mach number must occupy first place among the dimensionless parameters. Just as flow with viscosity depends strongly on the Reynolds number, and many geophysical flows depend strongly on the Rossby number, so high speed flow depends strongly on the Mach number. Since the compression induced by speed changes is negligible if $M_0^2 \ll 1$, high speed flow is defined as flow for which the square of the Mach number is not negligible compared with unity.

(b) Sound Waves

Consider a motionless nearly uniform fluid with typical density ρ_0 and sound speed c_0. The fluid is subject to small perturbations in pressure, density, and velocity. Linearization of the mass and momentum equations shows that the dominant balances include

$$\frac{\partial \rho}{\partial t} + \rho_0 \nabla \cdot \mathbf{u} \sim 0,$$

$$\rho_0 \frac{\partial \mathbf{u}}{\partial t} + \nabla p \sim 0. \tag{2.7.7}$$

Elimination of $\mathbf{u}$ gives

$$\frac{\partial^2 \rho}{\partial t^2} \sim \nabla^2 p. \tag{2.7.8}$$

Since $\Delta p / \Delta \rho \sim c_0^2$, we obtain

$$\frac{\partial^2 p}{\partial t^2} \sim c_0^2 \nabla^2 p,$$

$$\frac{\partial^2 \rho}{\partial t^2} \sim c_0^2 \nabla^2 \rho, \tag{2.7.9}$$

$$\frac{\partial^2}{\partial t^2} (\nabla \cdot \mathbf{u}) \sim c_0^2 \nabla^2 (\nabla \cdot \mathbf{u}).$$

Therefore perturbations in p, ρ, and $\nabla \cdot \mathbf{u}$ satisfy the wave equation and propagate at speed c_0. Such perturbations are sound waves and form the subject matter of acoustics. Much of the theory of high speed flow reduces to the theory of acoustics in the limit that amplitudes of pressure perturbations tend to zero. For example, the linear theory of thin supersonic aerofoils is part of the theory of acoustics and is often called the acoustic version of aerodynamics.

Consider a region of length scale l_1 in which there occur changes of flow on a time scale t_1. By the wave equation (2.7.9), the wavelength corresponding to a period t_1 is $c_0 t_1$. Therefore acoustic variations in the region are negligible if $l_1 \ll c_0 t_1$ (i.e., $l_1/c_0 \ll t_1$). But l_1/c_0 is the time taken for sound to cross the region. Therefore acoustic variations are negligible if this time is small compared with the time scale of variation of the flow. That is, in a region of given size, acoustic variations are important only if the flow is changing rapidly enough. For slower changes, the flow is effectively incompressible. In this book we consider acoustical phenomena only when their theory emerges naturally from the theory of high speed flow, as in the acoustic version of aerodynamics for example.

(c) Flow Affected by Scale Height

We now calculate the compressibility of fluid due to its own weight. Let the height coordinate, increasing upwards, be z, and let the unit vector pointing vertically upwards be $\mathbf{e}_z$. The acceleration due to gravity is g, and the body force per unit mass is $\mathbf{f} = -g\mathbf{e}_z$. Across a layer of depth l_0 we assume the fluid has pressure difference Δp, density difference $\Delta \rho$, and difference in specific volume Δv. We suppose that the momentum equation contains a dominant balance between $-\nabla p$ and $\rho \mathbf{f}$, that is, between ∇p and $\rho g \mathbf{e}_z$. Then

$$\frac{\Delta p}{l_0} \sim \rho g. \tag{2.7.10}$$

The typical sound speed is c_0, so that $\Delta p \sim c_0^2 \Delta \rho$. Eliminating Δp, and using $\rho v = 1$, we obtain

$$\frac{\Delta v}{v} \sim \frac{\Delta \rho}{\rho} \sim \frac{g l_0}{c_0^2}. \tag{2.7.11}$$

The condition for neglect of compressibility, $v^{-1} \Delta v \ll 1$, is therefore

$$l_0 \ll \frac{c_0^2}{g}. \tag{2.7.12}$$

The quantity c_0^2/g, with the dimensions of length, is called the scale height. For atmospheric air, the scale height is about 8.4 kilometers. Thus aerodynamic flows are unaffected by scale height; that is, in aerodynamics there are no significant changes in compression due to gravity.

2.8 Bibliographic Notes

The equations in this chapter are discussed in detail in many texts on fluid dynamics. See, for example, the chapters on physical properties of fluids and on the governing equations in Lighthill (1963) and Batchelor (1967); on ideal fluids in Milne–Thomson (1968); on gas dynamics in Whitham (1974); and on ideal fluids, viscosity, thermal conduction, and shock waves in Landau & Lifshitz (1987). See also the texts and monographs in Table 13.4.1. In this book we do not cover Kelvin's circulation theorem, which is treated at length in texts that emphasise incompressible flow.

3

Thermodynamics

3.1 General Laws and Definitions

Consider an infinitesimal fluid element with density ρ, specific volume v, pressure p, temperature T, and internal energy per unit mass e, also known as specific internal energy. We use the term "specific" to mean "per unit mass," but we omit it when no ambiguity arises. Other specific quantities defined later include enthalpy h, the Helmholtz function f, the Gibbs function g, and entropy s. The quantities ρ, v, $\dots$, s are thermodynamic variables as defined in Section 2.6, that is, they represent equilibrium values and satisfy equations of state. In general, any two of them can be varied independently, and the equations of state then determine the others. A simple exception is the pair ρ and v, related by $\rho = v^{-1}$; a less simple exception occurs at a phase boundary, for example between gas and liquid, at which there is a relation between p and T, the Clausius–Clapeyron equation.

If one of the thermodynamic variables is a known constant throughout a given region, then the value of a second thermodynamic variable, independent of the first, determines the values of the others by the equations of state. For example, if the entropy is constant, then any one of p, ρ, and T determines the other two, and we may write, for example, $p = p(\rho)$ or $p = p(T)$. In fluid dynamics, the relation between pressure and density is particularly important, and a flow in which the pressure depends only on the density (i.e., in which $p = p(\rho)$) is said to be barotropic. Many of the flows we shall consider are of this type, because we usually assume that upstream the fluid is in a state of uniform entropy (i.e., is homentropic) and further that the fluid is inviscid and non-heat-conducting. This model, which we discuss in Chapter 4, breaks down in the presence of shocks, but otherwise it provides a useful approximation to many types of high speed flow.

Suppose that an infinitesimal element of fluid undergoes a process of change from one equilibrium state to a neighbouring equilibrium state. The equilibrium

29

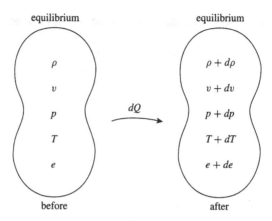

Fig. 3.1.1. Element of fluid before and after a change from equilibrium state ρ, v, p, T, e, ... to $\rho + d\rho$, $v + dv$, $p + dp$, $T + dT$, $e + de$, ... with heat flow dQ. The change may or may not be reversible (i.e., the intermediate states may or may not be in equilibrium).

values ρ, v, p, T, e, ... change to equilibrium values $\rho + d\rho$, $v + dv$, $p + dp$, $T + dT$, $e + de$, ... We consider a fluid element always consisting of the same fluid particles, not a small region fixed in space (i.e., we adopt a Lagrangian description). During the process, let the heat supplied per unit mass of the fluid element be dQ. The element of fluid before and after the heat addition is shown in Figure 3.1.1.

Although the changes of interest to us are from one equilibrium state to another, the intermediate states may or may not be in equilibrium. If a change takes place through a sequence of equilibrium states, it is called reversible. Consider a reversible change between states ρ, v, p, T, e, ... and $\rho + d\rho$, $v + dv$, $p + dp$, $T + dT$, $e + de$, ..., with heat dQ supplied per unit mass, and recall from Section 2.6 that we are measuring T in units of the kelvin (i.e., on the Kelvin temperature scale) so that T is the "absolute" or "thermodynamic" temperature. Then it follows from the second law of thermodynamics that $T^{-1} dQ$ is the differential of a well-defined property of the fluid, the specific entropy s. Thus

$$ds = \frac{dQ}{T}. \tag{3.1.1}$$

Therefore, the integral $\int T^{-1} dQ$ between any two equilibrium states, neighbouring or not, has the same value whatever the path taken, provided that every intermediate state is in equilibrium. Any one state may be chosen as a reference state, in which the entropy is s_0, and the entropy in any other state may be defined as the value of the integral $\int T^{-1} dQ$ along any suitable path which starts at the reference state and consists entirely of equilibrium states. If a change takes place at constant entropy, it is said to be isentropic.

A change from equilibrium state $\rho, \dots$ to equilibrium state $\rho + d\rho, \dots$ in an infinitesimal fluid element may take place without any heat flow across the boundary of the element, that is, so that $dQ = 0$, in which case the change is said to be adiabatic. Whether or not an adiabatic change is isentropic depends on whether the change is reversible; but a reversible adiabatic change is always isentropic. To see why, note that "reversible" implies $dQ = T ds$; "adiabatic" implies $dQ = 0$; and the third law of thermodynamics implies that $T > 0$. Therefore $ds = 0$, that is, the change is isentropic. An irreversible adiabatic change always produces an increase in entropy (i.e., leads to $ds > 0$), by the second law of thermodynamics. In many problems it is assumed from the outset that all changes are reversible, in which case the terms adiabatic and isentropic may be used interchangeably.

In a reversible change between two equilibrium states of a fluid element, each a state of rest, the heat supplied to the fluid element is $T ds$, and the work done on the surroundings by the boundary of the fluid element is $p dv$. Since the heat supplied equals the increase in internal energy plus the work done, it follows that

$$T ds = de + p dv. \tag{3.1.2}$$

This equation may be regarded as a combination of the first and second laws of thermodynamics. It is a relation between well-defined quantities for a fluid element in equilibrium (i.e., equilibrium thermodynamic variables), and so is not dependent for its validity on the assumption we made that the change from $s, e, v, \dots$ to $s + ds, e + de, v + dv, \dots$ is reversible. For example, suppose that the change takes place irreversibly, that is, that the intermediate states are not in equilibrium. Then the heat supplied is no longer $T ds$, and the work done on the surroundings is no longer $p dv$, but (3.1.2) remains true, as does the statement that the heat supplied equals the increase in internal energy plus the work done. By the same argument, that (3.1.2) contains only thermodynamic variables, it follows that (3.1.2) applies both to fluid at rest and to fluid in motion, provided that, if the distinction is maintained between equilibrium pressure and mechanical pressure, introduced in Section 2.5, the quantity p in (3.1.2) is the equilibrium pressure.

Note that local reversibility does not imply global reversibility. For example, consider a large region of fluid of nonuniform temperature, surrounded by a thermally insulating boundary, in which by the process of heat flowing from warmer to cooler regions the fluid is approaching a uniform temperature. In this process, illustrated in Figure 3.1.2, each fluid element may always be close to

3 Thermodynamics

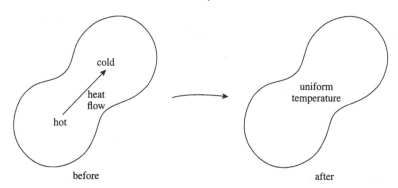

Fig. 3.1.2. Heat flow in a large region of fluid. The approach to uniform temperature is a globally irreversible change that can be described by locally reversible thermodynamics.

equilibrium, but the process as a whole is irreversible (i.e., the entropy of the fluid is increasing). Thus in the governing equations of fluid dynamics we use the equations of state for equilibrium thermodynamics, but we are nevertheless able to describe global irreversible changes.

We used the thermodynamic equation (3.1.2) in the sequence of equations (2.6.16)–(2.6.19) which led from the energy equation to an expression for $\rho T \, Ds/Dt$ in terms of shear viscosity μ, bulk viscosity ζ, and thermal conductivity k. For an ideal fluid, in which μ, ζ, and k may be taken to be zero, (2.6.19) shows that $Ds/Dt = 0$, as in the energy equation (2.6.24c). For a fluid which is not ideal, and for which no assumptions are made about heat flow at the boundary of the fluid or at the boundary of an infinitesimal fluid element, the value of Ds/Dt may be positive or negative. However, for a moving volume of nonideal fluid which contains always the same fluid particles and is thermally insulated from its surroundings, the second law of thermodynamics implies that the total entropy of the fluid cannot decrease. Thus if we denote the moving volume of fluid (i.e., the material region), by V, then

$$\frac{d}{dt} \int_V \rho \, s \, dV \geq 0. \tag{3.1.3}$$

After (2.6.22) we used (3.1.3) in deducing that μ, ζ, and k are always nonnegative.

The left-hand side of (3.1.3) may be rewritten by the standard result of vector calculus for material volumes to give

$$\int_V \rho \frac{Ds}{Dt} \, dV \geq 0. \tag{3.1.4}$$

Here no distinction need be made between a material and a fixed volume V, because the integral is not differentiated. The most common application of (3.1.4) is to flow through a shock wave, for which it implies $Ds/Dt \geq 0$, so that as fluid passes through the shock its entropy cannot decrease. Thus many solutions of the shock-wave jump conditions of mass, momentum, and energy can be shown not to correspond to a possible flow, because they would imply a physically impossible decrease in entropy. The fact that entropy increases at a shock was not at all obvious to early investigators.

The thermodynamic energy equation (3.1.2) is the starting point for some mathematical deductions based on differentials. We begin by rewriting (3.1.2) to give the differential of the internal energy e as

$$de = T ds - p dv. \tag{3.1.5}$$

Adding $d(pv)$ to each side converts $-p dv$ to $v dp$; subtracting $d(Ts)$ converts $T ds$ to $-s dT$; and adding $d(pv - Ts)$ converts $T ds - v dp$ to $-s dT + v dp$. Thus (3.1.5) is one of a set of four equivalent equations in e, h, f, and g where h, f, and g are defined by

$$h = e + pv,$$
$$f = e - Ts, \tag{3.1.6}$$
$$g = e + pv - Ts.$$

The four equivalent equations are (3.1.5) and

$$dh = d(e + pv) = T ds + v dp, \tag{3.1.7a}$$

$$df = d(e - Ts) = -s dT - p dv, \tag{3.1.7b}$$

$$dg = d(e + pv - Ts) = -s dT + v dp. \tag{3.1.7c}$$

Here h is the enthalpy, or heat function; f is the Helmholtz free energy, or available energy, or work function, or Helmholtz function; and g is the Gibbs free energy, or free enthalpy, or thermodynamic potential, or Gibbs function. The term thermodynamic potential is used in different senses: Sometimes all four of e, h, f, and g are called thermodynamic potentials, sometimes only f and g, and sometimes only g. The variables e, h, f, and g denote "specific" values, with physical dimensions of energy per unit mass (i.e., velocity squared).

We have assumed that two thermodynamic variables may be regarded as independent, and the other variables are then determined by equations of state. We shall indicate by a subscript the variable held constant during a partial differentiation. Then we obtain from (3.1.5) and (3.1.7), by equating mixed derivatives $\partial^2 e/\partial s \, \partial v = \partial^2 e/\partial v \, \partial s, \ldots$, the formulae

$$\left(\frac{\partial T}{\partial v}\right)_s = -\left(\frac{\partial p}{\partial s}\right)_v, \tag{3.1.8a}$$

$$\left(\frac{\partial T}{\partial p}\right)_s = \left(\frac{\partial v}{\partial s}\right)_p, \tag{3.1.8b}$$

$$\left(\frac{\partial s}{\partial v}\right)_T = \left(\frac{\partial p}{\partial T}\right)_v, \tag{3.1.8c}$$

$$\left(\frac{\partial s}{\partial p}\right)_T = -\left(\frac{\partial v}{\partial T}\right)_p. \tag{3.1.8d}$$

These are Maxwell's relations. The last two are especially useful when regarded as giving expressions for partial derivatives of s, because the right-hand sides are determined by the readily available (p, v, T) equation of state. For example, in conjunction with (3.1.5) and (3.1.7a) they give

$$\left(\frac{\partial e}{\partial v}\right)_T = T\left(\frac{\partial s}{\partial v}\right)_T - p = T\left(\frac{\partial p}{\partial T}\right)_v - p, \tag{3.1.9a}$$

$$\left(\frac{\partial h}{\partial p}\right)_T = T\left(\frac{\partial s}{\partial p}\right)_T + v = -T\left(\frac{\partial v}{\partial T}\right)_p + v. \tag{3.1.9b}$$

Two further thermodynamic variables are the specific heat at constant volume, c_v, and the specific heat at constant pressure, c_p, together known as the principal specific heats. They are defined as the heat supplied to the fluid per unit temperature rise per unit mass in a reversible change at constant volume and at constant pressure, respectively. The definitions imply the existence of two expressions for $T ds$, containing functions A and B, which we may specify later, of the form

$$T ds = c_v dT + A dv,$$
$$T ds = c_p dT + B dp. \tag{3.1.10}$$

To obtain these expressions, note that $T ds$ is a linear combination of the differentials of any two independent thermodynamic variables. The choice dT and dv is adapted to the definition of c_v, as seen by considering a reversible change with $dv = 0$; and the choice dT and dp is likewise adapted to the definition of c_p, as seen by considering a reversible change with $dp = 0$. Although this derivation of (3.1.10) is based on reversible changes, the same reasoning as used for (3.1.2) shows that (3.1.10) applies also to irreversible changes. For example, if a fluid element passes from one equilibrium state to another, but not via a sequence of equilibrium states, then (3.1.10) still applies, even though the heat supplied is not $T ds$. Equations (3.1.10) are equivalent to

$$c_v = T \left(\frac{\partial s}{\partial T} \right)_v = T \left(\frac{\partial s}{\partial T} \right)_\rho,$$

$$c_p = T \left(\frac{\partial s}{\partial T} \right)_p. \tag{3.1.11}$$

When $dv = 0$, we have $T ds = de$, by (3.1.5), and when $dp = 0$ we have $T ds = dh$, by (3.1.7a). Therefore (3.1.11) gives

$$c_v = \left(\frac{\partial e}{\partial T} \right)_v, \qquad c_p = \left(\frac{\partial h}{\partial T} \right)_p. \tag{3.1.12}$$

The ratio of specific heats γ is defined by

$$\gamma = \frac{c_p}{c_v}. \tag{3.1.13}$$

In general, c_p, c_v, and γ are functions of two independent thermodynamic variables. In Section 3.9 we consider the important special case in which c_p, c_v, and γ are functions only of T, and in Section 3.10 we consider the further special case, of even greater importance, in which they are all constants.

In (2.7.9) we used the fact that in acoustics the ratio $\Delta p / \Delta \rho$ of the pressure perturbations to the density perturbations is a measure of the square of the speed of sound. Texts on acoustics show that under normal conditions the thermodynamic quantities in a sound wave vary not at constant temperature as one might expect, and as was assumed by Newton, but at constant entropy (i.e., isentropically). Accordingly, we define the speed of sound c by

$$c^2 = \left(\frac{\partial p}{\partial \rho} \right)_s. \tag{3.1.14}$$

Sometimes it is convenient to regard on an equal footing the values of $\partial p / \partial \rho$ obtained when different variables are held fixed. Any value of $(\partial p / \partial \rho)^{\frac{1}{2}}$ obtained in this way is called "a speed of sound," and c as defined by (3.1.14) is then called the isentropic speed of sound.

Arbitrary changes $d\rho$ and ds produce a pressure change $dp = (\partial p / \partial \rho)_s \, d\rho + (\partial p / \partial s)_\rho \, ds$, that is,

$$dp = c^2 d\rho + \left(\frac{\partial p}{\partial s} \right)_\rho ds. \tag{3.1.15}$$

Isentropic changes are conveniently expressed in terms of formulae that contain c, since whenever $ds = 0$ we have $dp = c^2 d\rho$. Let us set $ds = 0$ in the relations $de = T ds - p dv$ and $dh = T ds + v dp$, remembering that $v = \rho^{-1}$ and $dv = -\rho^{-2} d\rho$. We obtain

$$\frac{d\rho}{\rho} = \frac{dp}{\rho c^2} = \frac{dh}{c^2} = \frac{\rho c^2}{p} \frac{de}{c^2}. \tag{3.1.16}$$

That is, in terms of the enthalpy change dh we have

$$\frac{dp}{\rho} = dh, \qquad \frac{d\rho}{\rho} = \frac{dh}{c^2}, \qquad de = \frac{p}{\rho c^2} dh. \tag{3.1.17}$$

These equations imply that, for nonisentropic changes, there are expressions, containing functions A_1, A_2, and A_3, which may be calculated if required, of the form

$$\frac{dp}{\rho} = dh + A_1 ds, \tag{3.1.18a}$$

$$\frac{d\rho}{\rho} = \frac{dh}{c^2} + A_2 ds, \tag{3.1.18b}$$

$$de = \frac{p}{\rho c^2} dh + A_3 ds. \tag{3.1.18c}$$

In thermodynamic formulae, the differential d may be replaced by other differential operators, such as the spatial gradient ∇ or the convective derivative D/Dt. If entropy is uniform over a region of space, so that $\nabla s = 0$, then (3.1.17) or (3.1.18) gives

$$\frac{\nabla p}{\rho} = \nabla h, \qquad \frac{\nabla \rho}{\rho} = \frac{\nabla h}{c^2}, \qquad \nabla e = \frac{p}{\rho c^2} \nabla h. \tag{3.1.19}$$

If entropy is constant following a fluid particle, so that $Ds/Dt = 0$, then similarly

$$\frac{1}{\rho} \frac{Dp}{Dt} = \frac{Dh}{Dt}, \qquad \frac{1}{\rho} \frac{D\rho}{Dt} = \frac{1}{c^2} \frac{Dh}{Dt}, \qquad \frac{De}{Dt} = \frac{p}{\rho c^2} \frac{Dh}{Dt}. \tag{3.1.20}$$

Compared with density and pressure, the specific enthalpy and specific entropy might seem a little recondite. However, they are often convenient variables in which to write the equations of motion, and we shall later make use of (3.1.17)–(3.1.20).

3.2 Practical Thermodynamic Formulae

A natural way of introducing thermodynamic quantities is to begin with an equation of state in the variables p, v, and T, and then define e, which makes possible the definitions of h, c_v, c_p, and γ. Then s is defined, and hence so are f, g, and c. Unfortunately, this order leads to an unsystematic presentation of results. The reason, evident in (3.1.5)–(3.1.7), is that p, v, T, and s form one group of variables, in which (p, v) and (T, s) are two pairs, and e, h, f, and g form another group of variables. The physical dimensions of pv, Ts, e, h, f, and g are those of $c_v T$ or $c_p T$ (i.e., energy per unit mass, or velocity squared).

In Sections 3.2–3.8 we derive a systematic and rather complete set of thermodynamic formulae, applicable to an arbitrary homogeneous fluid, in which all results are expressed in variables suitable for problems of high speed flow. These variables include the coefficient of thermal expansion β, the isothermal compressibility κ, and the coefficient of nonlinearity Γ, also called the fundamental gas-dynamic derivative. Their definitions are

$$\beta = \frac{1}{v}\left(\frac{\partial v}{\partial T}\right)_p, \tag{3.2.1a}$$

$$\kappa = -\frac{1}{v}\left(\frac{\partial v}{\partial p}\right)_T, \tag{3.2.1b}$$

$$\Gamma = \frac{1}{c}\left(\frac{\partial(\rho c)}{\partial \rho}\right)_s = 1 + \frac{\rho}{c}\left(\frac{\partial c}{\partial \rho}\right)_s = 1 + \frac{\rho}{2c^2}\left(\frac{\partial^2 p}{\partial \rho^2}\right)_s$$

$$= \frac{v^3}{2c^2}\left(\frac{\partial^2 p}{\partial v^2}\right)_s = \frac{c^4}{2v^3}\left(\frac{\partial^2 v}{\partial p^2}\right)_s. \tag{3.2.1c}$$

Note that $1/\kappa$ is the isothermal bulk modulus of elasticity of the fluid. Thermodynamic arguments show that $\kappa > 0$ for all substances. We usually assume that $\beta > 0$, since nearly all substances expand on heating at constant pressure; an exception is water below $4°C$, for which $\beta < 0$. A substance that satisfies the inequality $(\partial^2 p/\partial v^2)_s > 0$ is said to have a convex equation of state, since curves of fixed s in the (p, v) plane are then convex when viewed from below. Except near a critical point, nearly all substances have a convex equation of state, and so by (3.2.1c) we usually assume that $\Gamma > 0$. Nevertheless, phenomena in which $\Gamma < 0$ are of interest.

To make explicit the way in which our formulae simplify for a perfect gas, defined in Section 3.9, we use wherever possible the dimensionless quantities $p/(\rho c^2/\gamma)$ and βT, which for a perfect gas each take the value 1. In effect, we scale p with $\rho c^2/\gamma$ and T with $1/\beta$. The resulting formulae, in which we also use the dimensionless quantities s/c_v, $e/(c_v T)$, $h/(c_v T)$, $f/(c_v T)$, and

$g/(c_v T)$, are uniform in appearance and often the only parts not obtainable by inspection are powers of βT with coefficients that depend on γ.

The above quantities satisfy the identities

$$\frac{\rho c^2}{\gamma} = \frac{1}{\kappa} = \frac{(\gamma - 1)\rho c_v T}{(\beta T)^2}. \qquad (3.2.2\text{a,b})$$

Here each term has the dimensions of p, and for a perfect gas each term equals p. The identities (3.2.2) provide the key to expressing in intelligible variables nearly all of the thermodynamic formulae that arise in practice. We shall derive the identities later in this section, and for the moment determine some equivalent forms.

The outer terms in (3.2.2) give expressions for the dimensionless quantities $c^2/(c_p T)$ and $pv/(c_v T)$, that is,

$$\frac{c^2}{c_p T} = \frac{\gamma - 1}{(\beta T)^2}, \qquad \frac{pv}{c_v T} = \frac{\gamma - 1}{(\beta T)^2} \frac{p}{\rho c^2/\gamma}. \qquad (3.2.3)$$

Instead of $p/(\rho c^2/\gamma)$ we could more compactly write κp, analogously to βT, but it is convenient to retain ρ and c^2 explicitly. The relation $\rho = 1/v$, and the partial derivatives defining c^2 and κ, show that (3.2.2a) is equivalent to

$$\frac{(\partial p/\partial v)_s}{(\partial p/\partial v)_T} = \gamma. \qquad (3.2.4)$$

With $\gamma = c_p/c_v$, and c_p and c_v given by (3.1.11), this is

$$\frac{(\partial p/\partial v)_s}{(\partial p/\partial v)_T} = \frac{(\partial s/\partial T)_p}{(\partial s/\partial T)_v}. \qquad (3.2.5)$$

Equation (3.2.2b) is usually written

$$c_p - c_v = \frac{\beta^2 v T}{\kappa}. \qquad (3.2.6)$$

Insertion of the definitions of β and κ, followed by use of identity (3.2.10) below, applied to p, v, and T, shows that (3.2.6) is equivalent to

$$c_p - c_v = T \left(\frac{\partial v}{\partial T} \right)_p \left(\frac{\partial p}{\partial T} \right)_v. \qquad (3.2.7)$$

By Maxwell's relation for $(\partial p/\partial T)_v$, this is

$$c_p - c_v = T \left(\frac{\partial v}{\partial T} \right)_p \left(\frac{\partial s}{\partial v} \right)_T, \qquad (3.2.8)$$

or, from the expressions for c_p and c_v in (3.1.11),

$$\left(\frac{\partial s}{\partial T}\right)_p - \left(\frac{\partial s}{\partial T}\right)_v = \left(\frac{\partial v}{\partial T}\right)_p \left(\frac{\partial s}{\partial v}\right)_T. \qquad (3.2.9)$$

We now derive the above formulae. The mathematical identities on which they rest are, first, that if x, y, and z are functionally related, then

$$\left(\frac{\partial x}{\partial y}\right)_z \left(\frac{\partial y}{\partial z}\right)_x \left(\frac{\partial z}{\partial x}\right)_y = -1, \qquad (3.2.10)$$

and, second, that if x, y, z, and w are such that any three of them are functionally related, then

$$\left(\frac{\partial x}{\partial y}\right)_w \left(\frac{\partial y}{\partial z}\right)_w = \left(\frac{\partial x}{\partial z}\right)_w, \qquad (3.2.11\text{a})$$

$$\left(\frac{\partial x}{\partial y}\right)_w \left(\frac{\partial z}{\partial w}\right)_x = \left(\frac{\partial x}{\partial y}\right)_z \left(\frac{\partial z}{\partial w}\right)_y, \qquad (3.2.11\text{b})$$

$$\left(\frac{\partial w}{\partial z}\right)_x - \left(\frac{\partial w}{\partial z}\right)_y = \left(\frac{\partial y}{\partial z}\right)_x \left(\frac{\partial w}{\partial y}\right)_z = -\left(\frac{\partial x}{\partial z}\right)_y \left(\frac{\partial w}{\partial x}\right)_z. \qquad (3.2.11\text{c})$$

Of these identities, which are proved in texts on calculus, we note only that (3.2.11c) is part of the chain rule for a change of variables in w from $w(x, z)$ to $w(y, z)$, or vice versa. The edifice of formulae in classical thermodynamics, built on (3.2.2), amounts to consequences of (3.2.11b,c) applied to the variables $(x, y, z, w) = (p, v, T, s)$. Thus (3.2.5), which is equivalent to (3.2.2a), is an example of (3.2.11b); and (3.2.9), which is equivalent to (3.2.2b), is an example of (3.2.11c). Hence we have derived all the results in this section.

In what follows, Section 3.3 contains derivatives obtainable from p, v, and T alone but gives the results in terms of quantities such as γ and c^2, which are defined in terms of further variables; Section 3.4 contains derivatives of s and derivatives at fixed s; and Sections 3.5–3.8 are similar to Section 3.4 but contain formulae in e, h, f, and g. The results, which are valid for an arbitrary homogeneous fluid, are applied in Section 3.9 to a perfect gas by putting $\beta T = 1$ and $p = \rho c^2/\gamma$, and they are applied in Section 3.10 to a perfect gas with constant specific heats (i.e., to a polytropic gas). Since fluid-dynamical formulae are conveniently written in terms of ρ rather than v, and because our aim is to use familiar variables, we make somewhat irregular use of the substitutions $v = 1/\rho$ and $v\partial/\partial v = -\rho\partial/\partial\rho$. The result is a practical set of formulae for use in problems of high speed flow. As Sections 3.3–3.8 are in part a collection of formulae for reference purposes, readers may prefer to turn at once to the theory of a perfect gas in Section 3.9 and refer to Sections 3.3–3.8 only as necessary.

3.3 Pressure, Volume, and Temperature

The equation of state in p, v, and T determines differential relations in dp, dv, and dT, for example

$$dv = \left(\frac{\partial v}{\partial p}\right)_T dp + \left(\frac{\partial v}{\partial T}\right)_p dT. \tag{3.3.1}$$

By definition of β and κ, this is

$$\frac{dv}{v} = -\kappa dp + \beta dT. \tag{3.3.2}$$

Since $\kappa = (\rho c^2/\gamma)^{-1}$ and $\rho v = 1$, we may also write

$$\frac{dp}{\rho c^2/\gamma} = -\frac{dv}{v} + (\beta T)\frac{dT}{T} = \frac{d\rho}{\rho} + (\beta T)\frac{dT}{T}. \tag{3.3.3}$$

Equation (3.3.3) is in standard form, that is, it makes explicit the dimensionless groups βT and $dp/(\rho c^2/\gamma)$. It represents, compactly, the partial derivatives obtainable from (p, v, T) or (p, ρ, T), since these are ratios of the coefficients of the differentials. Thus

$$\frac{v}{\rho c^2/\gamma}\left(\frac{\partial p}{\partial v}\right)_T = -1, \qquad \frac{T}{\rho c^2/\gamma}\left(\frac{\partial p}{\partial T}\right)_v = \beta T, \qquad \frac{T}{v}\left(\frac{\partial v}{\partial T}\right)_p = \beta T.$$
$$\tag{3.3.4}$$

Although (3.3.4) is at first sight a rather cumbersome way of presenting the information contained in (3.3.2), the method we have here demonstrated, of reducing equations and derivatives to a standard form, will be found helpful in later sections.

3.4 Entropy

The main differential relations in s are, in standard form,

$$\frac{ds}{c_v} = \frac{1}{\beta T}\frac{dp}{\rho c^2/\gamma} + \frac{\gamma}{\beta T}\frac{dv}{v} \tag{3.4.1a}$$

$$= -\frac{\gamma - 1}{\beta T}\frac{dp}{\rho c^2/\gamma} + \gamma\frac{dT}{T} \tag{3.4.1b}$$

$$= \frac{\gamma - 1}{\beta T}\frac{dv}{v} + \frac{dT}{T}. \tag{3.4.1c}$$

These are obtained from (3.3.4) and such equations as $(\partial s/\partial p)_T = -(\partial v/\partial T)_p$, $(\partial s/\partial T)_p = c_p/T = \gamma c_v/T$, $(\partial s/\partial v)_T = (\partial p/\partial T)_v$, and $(\partial s/\partial T)_v = c_v/T$, which are consequences of Maxwell's relations and the definitions of specific heats. At any stage, (3.3.3) may be used to express any of dp, dv, and dT as a linear combination of the other two.

The coefficients in (3.4.1) give

$$\frac{\rho c^2/\gamma}{c_v}\left(\frac{\partial s}{\partial p}\right)_v = \frac{1}{\beta T}, \qquad \frac{v}{c_v}\left(\frac{\partial s}{\partial v}\right)_p = \frac{\gamma}{\beta T},$$

$$\frac{\rho c^2/\gamma}{c_v}\left(\frac{\partial s}{\partial p}\right)_T = -\frac{\gamma-1}{\beta T}, \qquad \frac{T}{c_v}\left(\frac{\partial s}{\partial T}\right)_p = \gamma, \qquad (3.4.2)$$

$$\frac{v}{c_v}\left(\frac{\partial s}{\partial v}\right)_T = \frac{\gamma-1}{\beta T}, \qquad \frac{T}{c_v}\left(\frac{\partial s}{\partial T}\right)_v = 1.$$

The derivatives of p, v, and T at fixed s, obtained from (3.4.1) by putting $ds = 0$, are, in standard form,

$$\frac{v}{\rho c^2/\gamma}\left(\frac{\partial p}{\partial v}\right)_s = -\gamma, \qquad \frac{T}{\rho c^2/\gamma}\left(\frac{\partial p}{\partial T}\right)_s = \left(\frac{\gamma}{\gamma-1}\right)\beta T,$$

$$\frac{T}{v}\left(\frac{\partial v}{\partial T}\right)_s = -\frac{\beta T}{\gamma-1}. \qquad (3.4.3)$$

These may be compared with the derivatives in (3.3.4). The ratios are

$$\frac{(\partial p/\partial v)_s}{(\partial p/\partial v)_T} = \gamma, \qquad \frac{(\partial p/\partial T)_s}{(\partial p/\partial T)_v} = \frac{\gamma}{\gamma-1}, \qquad \frac{(\partial v/\partial T)_s}{(\partial v/\partial T)_p} = -\frac{1}{\gamma-1}.$$

$$(3.4.4)$$

Useful thermodynamic variables obtained from (3.4.3) are the isentropic exponent m and the Grüneisen parameter G, defined by $m = -(v/p)(\partial p/\partial v)_s$ and $G = -(v/T)(\partial T/\partial v)_s$. Equation (3.4.3) and the identities (3.2.2) give

$$m = \frac{c^2}{pv} = \frac{\gamma}{\kappa p}, \qquad (3.4.5a)$$

$$G = \frac{\gamma-1}{\beta T} = \frac{\beta v}{\kappa c_v} = \frac{\beta c^2}{\gamma c_v} = \frac{\beta c^2}{c_p}. \qquad (3.4.5b)$$

The reason for the name isentropic exponent is that, at fixed s, if m is approximately constant then so is pv^m. Again at fixed s, if G is approximately constant then so is Tv^G. The Grüneisen parameter has the same sign as β, and so usually $G > 0$. Although $m = \gamma$ for a perfect gas, the two quantities m and γ are not equal in general, and it is therefore unfortunate that some writers use the symbol γ for the isentropic exponent of an arbitrary fluid.

The fluid-dynamical energy equation, (2.6.19), in which the left-hand side is $\rho T\, Ds/Dt$, may be rewritten using the equations we have just derived. For example, (3.4.1a) gives

$$\frac{Dp}{Dt} - c^2\frac{D\rho}{Dt} = \frac{\beta c^2}{c_p}\{2\mu\bar{\mathbf{e}}^2 + \zeta(\nabla\cdot\mathbf{u})^2 + \nabla\cdot(k\nabla T)\}. \tag{3.4.6}$$

Here the coefficient on the right is the Grüneisen parameter (3.4.5b).

Isentropic processes are so important in practice that it is convenient to use a standard form for higher derivatives at fixed s. This is achieved by using the coefficient of nonlinearity Γ, a dimensionless quantity, as defined in (3.2.1c). Thus

$$\frac{v^2}{\rho c^2/\gamma}\left(\frac{\partial^2 p}{\partial v^2}\right)_s = 2\gamma\Gamma, \qquad \frac{(\rho c^2/\gamma)^2}{v}\left(\frac{\partial^2 v}{\partial p^2}\right)_s = \frac{2\Gamma}{\gamma^2}. \tag{3.4.7}$$

If the first of these is expressed in terms of ρ, the result is

$$\frac{\rho^2}{\rho c^2/\gamma}\left(\frac{\partial^2 p}{\partial\rho^2}\right)_s = 2\gamma(\Gamma - 1). \tag{3.4.8}$$

Thus the sign of $(\partial^2 p/\partial\rho^2)_s$ is that of $\Gamma - 1$. Some derivatives of c in standard form are

$$\frac{v}{c}\left(\frac{\partial c}{\partial v}\right)_s = -(\Gamma - 1), \qquad \frac{\rho}{c}\left(\frac{\partial c}{\partial\rho}\right)_s = \Gamma - 1,$$

$$\frac{T}{c}\left(\frac{\partial c}{\partial T}\right)_s = \left(\frac{\Gamma - 1}{\gamma - 1}\right)\beta T. \tag{3.4.9}$$

3.5 Internal Energy

Derivatives of e may be obtained by substitution of expressions (3.4.1) for ds into the basic equation $de = T\,ds - p\,dv$. The resulting formulae contain the dimensionless quantity $pv/(c_v T)$, which may be written in standard form by (3.2.3). This gives

$$\begin{aligned}
\frac{de}{c_v T} &= \frac{1}{\beta T}\frac{dp}{\rho c^2/\gamma} + \left(\frac{\gamma}{\beta T} - \frac{\gamma - 1}{(\beta T)^2}\frac{p}{\rho c^2/\gamma}\right)\frac{dv}{v} \\
&= \left(-\frac{\gamma - 1}{\beta T} + \frac{\gamma - 1}{(\beta T)^2}\frac{p}{\rho c^2/\gamma}\right)\frac{dp}{\rho c^2/\gamma} + \left(\gamma - \frac{\gamma - 1}{\beta T}\frac{p}{\rho c^2/\gamma}\right)\frac{dT}{T} \\
&= \left(\frac{\gamma - 1}{\beta T} - \frac{\gamma - 1}{(\beta T)^2}\frac{p}{\rho c^2/\gamma}\right)\frac{dv}{v} + \frac{dT}{T}. \tag{3.5.1}
\end{aligned}$$

The coefficients give the derivatives of e and also the derivatives of p, v, and T at fixed e. One such derivative of theoretical importance is $(\partial T/\partial v)_e$, the Joule coefficient, which determines whether a temperature change occurs during an expansion at fixed e; it is zero for a perfect gas. The last of Equations (3.5.1) shows that, in standard form,

$$\frac{v}{T}\left(\frac{\partial T}{\partial v}\right)_e = -\frac{\gamma-1}{\beta T} + \frac{\gamma-1}{(\beta T)^2}\frac{p}{\rho c^2/\gamma}. \tag{3.5.2}$$

The isentropic derivatives of e are

$$\frac{\rho c^2/\gamma}{c_v T}\left(\frac{\partial e}{\partial p}\right)_s = \frac{\gamma-1}{\gamma}\frac{1}{(\beta T)^2}\frac{p}{\rho c^2/\gamma},$$

$$\frac{v}{c_v T}\left(\frac{\partial e}{\partial v}\right)_s = -\frac{\gamma-1}{(\beta T)^2}\frac{p}{\rho c^2/\gamma}, \tag{3.5.3}$$

$$\frac{1}{c_v}\left(\frac{\partial e}{\partial T}\right)_s = \frac{1}{\beta T}\frac{p}{\rho c^2/\gamma}.$$

These are obtained from $de = T\,ds - p\,dv$ and the intermediate steps $(\partial e/\partial p)_s = -p(\partial v/\partial p)_s$, $(\partial e/\partial v)_s = -p$, and $(\partial e/\partial T)_s = -p(\partial v/\partial T)_s$.

3.6 Enthalpy

Derivatives of h are obtained as above, but starting from $dh = T\,ds + v\,dp$. The result is

$$\begin{aligned}
\frac{dh}{c_v T} &= \left(\frac{1}{\beta T} + \frac{\gamma-1}{(\beta T)^2}\right)\frac{dp}{\rho c^2/\gamma} + \frac{\gamma}{\beta T}\frac{dv}{v} \\
&= \left(-\frac{\gamma-1}{\beta T} + \frac{\gamma-1}{(\beta T)^2}\right)\frac{dp}{\rho c^2/\gamma} + \gamma\frac{dT}{T} \\
&= \left(\frac{\gamma-1}{\beta T} - \frac{\gamma-1}{(\beta T)^2}\right)\frac{dv}{v} + \left(1 + \frac{\gamma-1}{\beta T}\right)\frac{dT}{T}.
\end{aligned} \tag{3.6.1}$$

The coefficients give the derivatives of h and also the derivatives of p, v, and T at fixed h. Of practical importance in achieving low temperatures is $(\partial T/\partial p)_h$, the Joule–Kelvin coefficient, or Joule–Thomson coefficient, which determines the temperature change on expansion, at fixed h, into a region of lower pressure; it is zero for a perfect gas. The second of Equations (3.6.1) shows that

$$\frac{\rho c^2/\gamma}{T}\left(\frac{\partial T}{\partial p}\right)_h = \frac{1}{\gamma}\left(\frac{\gamma-1}{\beta T} - \frac{\gamma-1}{(\beta T)^2}\right). \tag{3.6.2}$$

Thus the Joule–Kelvin coefficient is zero when $\beta T = 1$, which determines a curve in the (p, T) plane, known as the inversion curve, separating the region of cooling from the region of heating.

The isentropic derivatives of h are

$$\frac{\rho c^2/\gamma}{c_v T} \left(\frac{\partial h}{\partial p} \right)_s = \frac{\gamma - 1}{(\beta T)^2},$$

$$\frac{v}{c_v T} \left(\frac{\partial h}{\partial v} \right)_s = -\frac{\gamma(\gamma - 1)}{(\beta T)^2}, \qquad (3.6.3)$$

$$\frac{1}{c_v} \left(\frac{\partial h}{\partial T} \right)_s = \frac{\gamma}{\beta T}.$$

These follow from $dh = T\,ds + v\,dp$ and the intermediate steps $(\partial h/\partial p)_s = v$, $(\partial h/\partial v)_s = v(\partial p/\partial v)_s$, and $(\partial h/\partial T)_s = v(\partial p/\partial T)_s$.

3.7 The Helmholtz Function

Starting from $df = -s\,dT - p\,dv$, and proceeding as above, we obtain

$$\frac{df}{c_v T} = -\frac{1}{\beta T} \frac{s}{c_v} \frac{dp}{\rho c^2/\gamma} - \left(\frac{1}{\beta T} \frac{s}{c_v} + \frac{\gamma - 1}{(\beta T)^2} \frac{p}{\rho c^2/\gamma} \right) \frac{dv}{v}$$

$$= \frac{\gamma - 1}{(\beta T)^2} \frac{p}{\rho c^2/\gamma} \frac{dp}{\rho c^2/\gamma} - \left(\frac{s}{c_v} + \frac{\gamma - 1}{\beta T} \frac{p}{\rho c^2/\gamma} \right) \frac{dT}{T} \qquad (3.7.1)$$

$$= -\frac{\gamma - 1}{(\beta T)^2} \frac{p}{\rho c^2/\gamma} \frac{dv}{v} - \frac{s}{c_v} \frac{dT}{T}.$$

The coefficients give the derivatives of f and also the derivatives of p, v, and T at fixed f.

The isentropic derivatives of f are

$$\frac{\rho c^2/\gamma}{c_v T} \left(\frac{\partial f}{\partial p} \right)_s = -\frac{\gamma - 1}{\gamma} \frac{1}{\beta T} \frac{s}{c_v} + \frac{\gamma - 1}{\gamma} \frac{1}{(\beta T)^2} \frac{p}{\rho c^2/\gamma},$$

$$\frac{v}{c_v T} \left(\frac{\partial f}{\partial v} \right)_s = \frac{\gamma - 1}{\beta T} \frac{s}{c_v} - \frac{\gamma - 1}{(\beta T)^2} \frac{p}{\rho c^2/\gamma}, \qquad (3.7.2)$$

$$\frac{1}{c_v} \left(\frac{\partial f}{\partial T} \right)_s = -\frac{s}{c_v} + \frac{1}{\beta T} \frac{p}{\rho c^2/\gamma}.$$

These follow from $df = -s\,dT - p\,dv$ and the intermediate steps $(\partial f/\partial p)_s = -s(\partial T/\partial p)_s - p(\partial v/\partial p)_s$, $(\partial f/\partial v)_s = -s(\partial T/\partial v)_s - p$, and $(\partial f/\partial T)_s = -s - p(\partial v/\partial T)_s$.

3.8 The Gibbs Function

Starting from $dg = -s\,dT + v\,dp$, we obtain

$$
\begin{aligned}
\frac{dg}{c_v T} &= \left(-\frac{s}{c_v} + \frac{\gamma - 1}{(\beta T)^2}\right)\frac{dp}{\rho c^2/\gamma} - \frac{1}{\beta T}\frac{s}{c_v}\frac{dv}{v} \\
&= \frac{\gamma - 1}{(\beta T)^2}\frac{dp}{\rho c^2/\gamma} - \frac{s}{c_v}\frac{dT}{T} \\
&= -\frac{\gamma - 1}{(\beta T)^2}\frac{dv}{v} + \left(-\frac{s}{c_v} + \frac{\gamma - 1}{\beta T}\right)\frac{dT}{T}.
\end{aligned}
\tag{3.8.1}
$$

The coefficients give the derivatives of g and also the derivatives of p, v, and T at fixed g.

The isentropic derivatives of g are

$$
\begin{aligned}
\frac{\rho c^2/\gamma}{c_v T}\left(\frac{\partial g}{\partial p}\right)_s &= -\frac{\gamma - 1}{\gamma}\frac{1}{\beta T}\frac{s}{c_v} + \frac{\gamma - 1}{(\beta T)^2}, \\
\frac{v}{c_v T}\left(\frac{\partial g}{\partial v}\right)_s &= \frac{\gamma - 1}{\beta T}\frac{s}{c_v} - \frac{\gamma(\gamma - 1)}{(\beta T)^2}, \\
\frac{1}{c_v}\left(\frac{\partial g}{\partial T}\right)_s &= -\frac{s}{c_v} + \frac{\gamma}{\beta T}.
\end{aligned}
\tag{3.8.2}
$$

These follow from $dg = -s\,dT + v\,dp$ and the intermediate steps $(\partial g/\partial p)_s = -s(\partial T/\partial p)_s + v$, $(\partial g/\partial v)_s = -s(\partial T/\partial v)_s + v(\partial p/\partial v)_s$, and $(\partial g/\partial T)_s = -s + v(\partial p/\partial T)_s$.

3.9 The Perfect Gas

A gas for which the (p, v, T) equation of state is $pv/T = $ constant is called a *perfect gas* or an *ideal gas*. We shall use the former term. The value of the constant, denoted R, varies from one perfect gas to another, but it is simply related to the molecular weight of the gas and a constant R_0 known as the universal gas constant, as we shall see below. Since gases are in most circumstances approximately perfect at pressures and densities that are not too high, the theoretical properties of a perfect gas are of great practical importance. Accordingly, we now determine in detail the consequences of the definitions and results in Sections 3.1–3.8 when they are supplemented by the single equation

$$
pv = RT.
\tag{3.9.1}
$$

In differential form, this is

$$\frac{dp}{p} + \frac{dv}{v} = \frac{dT}{T}.$$
(3.9.2)

A perfect gas is not the same as an ideal fluid, which is a hypothetical fluid without viscosity or thermal conductivity. Our terminology is standard among fluid dynamicists and many physicists, but not among engineers, who use the term "ideal gas" for a fluid that obeys (3.9.1).

On using (3.9.1) in (3.1.9), we obtain at once

$$\left(\frac{\partial e}{\partial v}\right)_T = T\left(\frac{\partial p}{\partial T}\right)_v - p = 0,$$

$$\left(\frac{\partial h}{\partial p}\right)_T = -T\left(\frac{\partial v}{\partial T}\right)_p + v = 0.$$
(3.9.3)

Therefore e and h depend only on T, and we may write

$$e = e(T), \qquad h = h(T).$$
(3.9.4)

We have thus obtained two of the most important properties of a perfect gas: that the specific internal energy and the specific enthalpy depend only on temperature. Either of these properties implies the other, since the definition $h = e + pv$, when applied to a perfect gas, gives

$$h = e + RT.$$
(3.9.5)

It is of interest to reverse the above steps and show that the equation of state of a perfect gas is a consequence of more basic assumptions. Consider a fluid for which (a) the internal energy depends only on temperature and (b) at fixed temperature, the equation of state tends to $pv = $ constant as $p \to 0$. Assumption (b) is Boyle's law. By (a), the left-hand side of (3.1.9a) vanishes, so that the right-hand side gives $(\partial p/\partial T)_v = p/T$, whence for some function $a(v)$ the equation of state of the fluid must be $p = a(v)T$. We cannot proceed further by (a) alone, but (b) gives $a(v) \propto 1/v$, and we recover (3.9.1). Thus a perfect gas may alternatively be defined as a fluid satisfying (a) and (b).

One might ask whether assumptions (a) and (b), or the equation of state $pv = RT$, are consequences of still simpler assumptions about the fluid. In thermodynamics proper, in which use is made only of macroscopic quantities, the answer is no. But if we are willing to make assumptions about microscopic structure and forces, then we have available the techniques of kinetic theory, in which the laws of dynamics are applied to individual molecules and their collisions, and macroscopic quantities are determined by averages. A perfect gas may then be defined as a gas in which the molecules have negligible volume and, except during collisions, negligible forces of interaction. This definition, according to which the internal energy is the sum of the energies of the individual molecules, may be shown to lead to the equation of state $pv = (k/\bar{m})T$, where $\bar{m}$ is the average mass of the molecules and k is Boltzmann's constant. Thus we recover (3.9.1) with $R = k/\bar{m}$. It is conventional to nondimensionalise $\bar{m}$ by dividing by one twelfth of the mass m_{12} of a carbon atom, to give the average molecular weight $\bar{M} = 12\bar{m}/m_{12}$, a dimensionless quantity, which in terms of Avogadro's constant N_A, defined by $N_A = 12/m_{12}$, is $\bar{M} = N_A\bar{m}$. Hence in terms of the universal gas constant R_0, defined by $R_0 = kN_A$, we have $R = R_0/\bar{M}$, and the equation of state for a perfect gas is $pv = (R_0/\bar{M})T$. In calculations, a convenient unit of mass is often the kilogram-mole, or molar mass, defined as $\bar{M}$ kilograms. A mole itself is a dimensionless quantity, and the term "mole" in an expression such as kilogram-mole may be thought of as representing the number $\bar{M}$. In published algebraic formulae expressed in moles, some terms called "numbers" at first sight appear not to be dimensionless.

Returning now to the fact that in a perfect gas e and h depend only on T, we find from the relations $c_v = (\partial e/\partial T)_v$ and $c_p = (\partial h/\partial T)_p$ given in (3.1.12) that c_v and c_p likewise depend only on T, that is,

$$c_v = c_v(T) = e'(T),$$
$$c_p = c_p(T) = h'(T). \tag{3.9.6}$$

Differentiation of (3.9.5) then gives

$$c_p(T) = c_v(T) + R. \tag{3.9.7}$$

This is Carnot's law. The ratio of specific heats, $\gamma = c_p/c_v$, also depends only on T, that is,

$$c_p(T) = \gamma(T)c_v(T). \tag{3.9.8}$$

The last two equations give

$$c_v(T) = \frac{R}{\gamma(T) - 1}, \qquad c_p(T) = \frac{\gamma(T)R}{\gamma(T) - 1}. \tag{3.9.9}$$

Therefore

$$e(T) = \int c_v(T)\,dT = R \int \frac{dT}{\gamma(T) - 1}, \tag{3.9.10a}$$

$$h(T) = \int c_p(T)\,dT = R \int \frac{\gamma(T)\,dT}{\gamma(T) - 1}. \tag{3.9.10b}$$

In conjunction with the above equations, especially (3.9.2), the general relation $T ds = de + p dv$ gives

$$\frac{ds}{c_v(T)} = \frac{dp}{p} + \gamma(T)\frac{dv}{v} \qquad (3.9.11a)$$

$$= -(\gamma(T) - 1)\frac{dp}{p} + \gamma(T)\frac{dT}{T} \qquad (3.9.11b)$$

$$= (\gamma(T) - 1)\frac{dv}{v} + \frac{dT}{T}. \qquad (3.9.11c)$$

Therefore the isentropic derivatives are

$$\left(\frac{\partial p}{\partial v}\right)_s = -\frac{\gamma(T)p}{v}, \quad \left(\frac{\partial p}{\partial T}\right)_s = \frac{\gamma(T)}{\gamma(T) - 1}\frac{p}{T}, \quad \left(\frac{\partial v}{\partial T}\right)_s = \frac{-1}{\gamma(T) - 1}\frac{v}{T}.$$
$$(3.9.12)$$

On multiplying (3.9.11) by $c_v(T)$, and integrating from a reference state (p_0, v_0, T_0), for which $p_0 v_0 = R T_0$ and $s = s_0$, we obtain

$$s - s_0 = \int_{(p_0, v_0)}^{(p,v)} c_v(T)\frac{dp}{p} + c_p(T)\frac{dv}{v} = -R \ln\left(\frac{p}{p_0}\right) + \int_{T_0}^{T} c_p(T)\frac{dT}{T}$$

$$= R \ln\left(\frac{v}{v_0}\right) + \int_{T_0}^{T} c_v(T)\frac{dT}{T}. \qquad (3.9.13)$$

Therefore

$$\frac{s - s_0}{R} = \int_{(p_0, v_0)}^{(p,v)} \frac{1}{\gamma(T) - 1}\frac{dp}{p} + \frac{\gamma(T)}{\gamma(T) - 1}\frac{dv}{v}$$

$$= -\ln\left(\frac{p}{p_0}\right) + \int_{T_0}^{T} \frac{\gamma(T)}{\gamma(T) - 1}\frac{dT}{T} = \ln\left(\frac{v}{v_0}\right) + \int_{T_0}^{T} \frac{1}{\gamma(T) - 1}\frac{dT}{T}.$$
$$(3.9.14)$$

In exponential form,

$$\exp\left(\frac{s - s_0}{R}\right) = \exp\left(\int_{(p_0, v_0)}^{(p,v)} \frac{1}{\gamma(T) - 1}\frac{dp}{p} + \frac{\gamma(T)}{\gamma(T) - 1}\frac{dv}{v}\right)$$

$$= \left(\frac{p}{p_0}\right)^{-1} \exp\left(\int_{T_0}^{T} \frac{\gamma(T)}{\gamma(T) - 1}\frac{dT}{T}\right)$$

$$= \left(\frac{v}{v_0}\right) \exp\left(\int_{T_0}^{T} \frac{1}{\gamma(T) - 1}\frac{dT}{T}\right). \qquad (3.9.15)$$

Hence at fixed entropy,

$$p \propto \exp\left(\int_{T_0}^{T} \frac{\gamma(T)}{\gamma(T)-1} \frac{dT}{T}\right),$$

$$v \propto \exp\left(-\int_{T_0}^{T} \frac{1}{\gamma(T)-1} \frac{dT}{T}\right).$$

(3.9.16)

Let e, h, f, and g take the values e_0, h_0, f_0, and g_0 at the reference state (p_0, v_0, T_0). The definition of f gives $f - f_0 = e - e_0 - (Ts - T_0 s_0)$, that is,

$$\frac{f - f_0}{RT} = \frac{e - e_0}{RT} - \frac{s - s_0}{R} - \frac{(T - T_0)}{T}\frac{s_0}{R}.$$

(3.9.17)

Hence (3.9.10a) and (3.9.14) give expressions for $(f - f_0)/(RT)$. The result may be written

$$\exp\left(\frac{f - f_0}{RT}\right) = \exp\left(\frac{1}{T}\int_{T_0}^{T} \frac{dT}{\gamma(T)-1} - \frac{T - T_0}{T}\frac{s_0}{R}\right)\exp\left(-\frac{s - s_0}{R}\right),$$

(3.9.18)

with the last term given by the reciprocal of any of the right-hand sides in (3.9.15). Similarly, the definition of g gives $g - g_0 = h - h_0 - (Ts - T_0 s_0)$, that is,

$$\frac{g - g_0}{RT} = \frac{h - h_0}{RT} - \frac{s - s_0}{R} - \frac{T - T_0}{T}\frac{s_0}{R}.$$

(3.9.19)

Hence (3.9.10b) and (3.9.14) give expressions for $(g - g_0)/(RT)$. Analogously to (3.9.18), we obtain

$$\exp\left(\frac{g - g_0}{RT}\right) = \exp\left(\frac{1}{T}\int_{T_0}^{T} \frac{\gamma(T)dT}{\gamma(T)-1} - \frac{T - T_0}{T}\frac{s_0}{R}\right)\exp\left(-\frac{s - s_0}{R}\right).$$

(3.9.20)

The results for an arbitrary fluid given in Sections 3.2–3.8 are expressed in terms of $c^2, \beta, \kappa, \Gamma, m$, and G, which for a perfect gas take the values

$$c^2 = \frac{\gamma(T)p}{\rho} = \gamma(T)RT, \qquad \beta = \frac{1}{T}, \qquad \kappa = \frac{1}{p},$$

$$\Gamma = \frac{1}{2}(\gamma(T)+1), \qquad m = \gamma(T), \qquad G = \gamma(T) - 1.$$

(3.9.21)

Note that c^2, β, Γ, m, and G depend only on T. The expression for c^2 is a consequence of (3.9.11a), and the expression for Γ may be obtained from

(3.2.1c) by the sequence of manipulations

$$\Gamma = 1 + \frac{\rho}{c}\left(\frac{\partial c}{\partial \rho}\right)_s = 1 - \frac{v}{c}\left(\frac{\partial c}{\partial v}\right)_s$$

$$= 1 - \frac{v}{c}\left(\frac{\partial c}{\partial T}\right)_s\left(\frac{\partial T}{\partial v}\right)_s$$

$$= 1 - \frac{v}{c}\frac{c}{2T}\left(-\frac{(\gamma(T)-1)T}{v}\right)$$ \hfill (3.9.22)

$$= \frac{1}{2}(\gamma(T) + 1).$$

Hence for a perfect gas the values of the various expressions appearing in Sections 3.2–3.8 are, with $\gamma = \gamma(T)$,

$$\beta T = 1, \qquad \kappa p = 1, \qquad \frac{p}{\rho c^2/\gamma} = 1,$$

$$\frac{c^2}{c_p T} = \gamma - 1, \qquad \frac{pv}{c_v T} = \gamma - 1, \qquad \frac{\beta^2 vT}{\kappa} = R,$$

$$\left(\frac{\partial^2 p}{\partial v^2}\right)_s = \frac{\gamma(\gamma+1)p}{v^2}, \qquad \left(\frac{\partial^2 v}{\partial p^2}\right)_s = \frac{\gamma+1}{\gamma^2}\frac{v}{p^2},$$ \hfill (3.9.23)

$$\left(\frac{\partial^2 p}{\partial \rho^2}\right)_s = \frac{\gamma(\gamma-1)p}{\rho^2}, \qquad \left(\frac{\partial c}{\partial v}\right)_s = -\frac{\gamma-1}{2}\frac{c}{v},$$

$$\left(\frac{\partial c}{\partial \rho}\right)_s = \frac{\gamma-1}{2}\frac{c}{\rho}, \qquad \left(\frac{\partial c}{\partial T}\right)_s = \frac{c}{2T}.$$

In the fluid-dynamical energy equation (3.4.6), the value of $\beta c^2/c_p$ is $\gamma(T) - 1$.

Formulae for a perfect gas can be obtained from the corresponding formulae in Sections 3.2–3.8 for an arbitrary fluid by putting $\beta T = 1$ and $p/(\rho c^2/\gamma) = 1$. Conversely, this fact has determined our choice of variables in Sections 3.2–3.8. The effect of a departure from the perfect-gas law may be assessed from the values of $\beta T - 1$ and $p/(\rho c^2/\gamma) - 1$, that is, $\kappa p - 1$.

3.10 The Perfect Gas with Constant Specific Heats

Over wide temperature ranges, a perfect gas with simple molecules has constant specific heats. This is because over these temperature ranges the molecules of the gas have a fixed number n of excited degrees of freedom, each of which makes a contribution $\frac{1}{2}RT$ to the specific internal energy e, so that, if the reference value of e is chosen so that $e = 0$ when $T = 0$, then

$$e = \tfrac{1}{2}nRT. \qquad (3.10.1)$$

Thus (3.9.6)–(3.9.8) give

$$c_v = \frac{de}{dT} = \frac{1}{2}nR, \qquad (3.10.2\text{a})$$

$$c_p = c_v + R = \left(\tfrac{1}{2}n + 1\right)R, \qquad (3.10.2\text{b})$$

$$\gamma = \frac{c_p}{c_v} = 1 + \frac{2}{n}. \qquad (3.10.2\text{c})$$

We noted in Section 3.9 that $R = k/\bar{m}$, where k is Boltzmann's constant and $\bar{m}$ is the average mass of the molecules of the gas. Thus $\tfrac{1}{2}RT = (\tfrac{1}{2}kT)/\bar{m}$. That is, the amount of energy $\tfrac{1}{2}RT$ per unit mass per excited degree of freedom is a consequence of the law of equipartition of energy, which asserts that the average amount of energy per molecule per excited degree of freedom is $\tfrac{1}{2}kT$.

At normal temperatures, a monatomic gas, such as helium, neon, or argon, has $n = 3$, corresponding to the translational energy of the motion of the centre of mass of a molecule in three coordinate directions; and a diatomic gas, such as hydrogen, nitrogen, oxygen, or carbon monoxide, has $n = 5$, corresponding to the further energy in the rotation of a molecule about two principal axes. Diatomic molecules do not store significant amounts of vibrational energy except at high temperatures, typically over several thousand degrees, although the small amount of vibrational energy in molecules of nitrogen and oxygen plays a key role in atmospheric sound attenuation at audio frequencies. We shall not consider ionization, which typically occurs at still higher temperatures, nor chemical reaction and dissociation. Thus we most frequently require (3.10.2c) with $n = 3$ or 5, that is,

$$\gamma = \begin{cases} \tfrac{5}{3} = 1.667 & \text{(monatomic gas)} \\[2mm] \tfrac{7}{5} = 1.4 & \text{(diatomic gas)}. \end{cases} \qquad (3.10.3)$$

Air, composed largely of nitrogen and oxygen, may usually be regarded as a diatomic gas with $\gamma = 1.4$, although the measured value is $\gamma = 1.401$ at 15°C and 1 atmosphere.

A perfect gas with constant specific heats is called a polytropic gas. The expressions for c_v and c_p in (3.9.9) show that for a perfect gas to be polytropic it is sufficient that γ is constant. Mathematically, the polytropic gases form a two-parameter family, labelled by parameters R and γ, for which all the integrals in Section 3.9 can be evaluated explicitly. Since real gases are closely polytropic over wide temperature ranges, the polytropic gases are a natural choice for presenting the theory of high speed flow. Often, the dependence of the results on γ is a matter of great practical importance, since this may determine the appropriate fluid to use in an experiment or an engine or wind tunnel.

The above equations give

$$\frac{1}{2}n = \frac{1}{\gamma - 1}, \qquad c_v = \frac{R}{\gamma - 1}, \qquad c_p = \frac{\gamma R}{\gamma - 1}. \tag{3.10.4}$$

Hence (3.10.1), together with the equation of state $pv = RT$ and the definition $h = e + pv$, gives

$$e = \frac{RT}{\gamma - 1}, \qquad h = \frac{\gamma RT}{\gamma - 1}. \tag{3.10.5}$$

At the reference state (p_0, v_0, T_0), for which $p_0 v_0 = RT_0$ and $s = s_0$, let e, h, f, and g take the values e_0, h_0, f_0, and g_0. Evaluation of the integrals in (3.9.13)–(3.9.15), with $\gamma(T) = \gamma = $ constant, gives

$$\exp\left(\frac{s - s_0}{c_v}\right) = \frac{p}{p_0}\left(\frac{v}{v_0}\right)^{\gamma} = \left(\frac{p}{p_0}\right)^{-(\gamma-1)}\left(\frac{T}{T_0}\right)^{\gamma} = \left(\frac{v}{v_0}\right)^{\gamma-1}\frac{T}{T_0},$$

$$\tag{3.10.6}$$

that is,

$$\exp\left(\frac{s}{c_v}\right) \propto \frac{p}{\rho^{\gamma}} \propto \frac{T^{\gamma}}{p^{\gamma-1}} \propto \frac{T}{\rho^{\gamma-1}}. \tag{3.10.7}$$

Similarly, (3.9.18) and (3.9.20) give

$$\exp\left(\frac{f - f_0}{c_v T}\right) = \exp\left(\frac{T - T_0}{T}\left(1 - \frac{s_0}{c_v}\right)\right)\exp\left(-\frac{s - s_0}{c_v}\right),$$

$$\exp\left(\frac{g - g_0}{c_v T}\right) = \exp\left(\gamma\left(\frac{T - T_0}{T}\right)\left(1 - \frac{s_0}{c_v}\right)\right)\exp\left(-\frac{s - s_0}{c_v}\right). \tag{3.10.8}$$

The terms in s are the reciprocals of the terms in (3.10.6). We have used c_v in denominators in (3.10.6) and (3.10.8) but R in (3.9.14)–(3.9.20). Equations (3.9.21)–(3.9.23) are available, now with constant γ, especially (3.9.21), which gives

$$c^2 = \frac{\gamma p}{\rho} = \gamma RT = \gamma(\gamma - 1)e = (\gamma - 1)h, \qquad (3.10.9a)$$

$$\beta = \frac{1}{T}, \qquad \kappa = \frac{1}{p}, \qquad \Gamma = \frac{1}{2}(\gamma + 1), \qquad m = \gamma, \qquad G = \gamma - 1.$$

$$(3.10.9b)$$

In differentials, (3.10.9a) is

$$2\frac{dc}{c} = \frac{dp}{p} - \frac{d\rho}{\rho} = \frac{dT}{T} = \frac{de}{e} = \frac{dh}{h}. \qquad (3.10.10)$$

The fluid-dynamical energy equation (3.4.6) becomes

$$\frac{Dp}{Dt} - c^2\frac{D\rho}{Dt} = (\gamma - 1)\{2\mu\tilde{e}^2 + \zeta(\nabla \cdot \mathbf{u})^2 + \nabla \cdot (k\nabla T)\}. \qquad (3.10.11)$$

At fixed s, a flow is barotropic (i.e., p is a function only of ρ), and any one of p, ρ, T, e, h, and c determines the others. Then (3.10.5), (3.10.7), and (3.10.9a) give

$$p \propto \rho^\gamma \propto T^{\gamma/(\gamma-1)} \propto e^{\gamma/(\gamma-1)} \propto h^{\gamma/(\gamma-1)} \propto c^{2\gamma/(\gamma-1)}, \qquad (3.10.12)$$

that is,

$$\frac{dp}{p} = \gamma\frac{d\rho}{\rho} = \frac{\gamma}{\gamma-1}\frac{dT}{T} = \frac{\gamma}{\gamma-1}\frac{de}{e} = \frac{\gamma}{\gamma-1}\frac{dh}{h} = \frac{2\gamma}{\gamma-1}\frac{dc}{c}. \qquad (3.10.13)$$

For air, taken to be a polytropic gas with $\gamma = 7/5$, the equations above give

$$c_v = \frac{5}{2}R, \qquad c_p = \frac{7}{2}R, \qquad e = \frac{5}{2}RT, \qquad h = \frac{7}{2}RT, \qquad c^2 = \frac{7}{5}RT.$$

$$(3.10.14)$$

The isentropic relations (3.10.12) give

$$p^{\frac{1}{7}} \propto \rho^{\frac{1}{5}} \propto T^{\frac{1}{2}} \propto e^{\frac{1}{2}} \propto h^{\frac{1}{2}} \propto c. \qquad (3.10.15)$$

Therefore in air at constant entropy, the speed of sound is proportional to the one-seventh power of the pressure and to the one-fifth power of the density. The proportionality to the square root of temperature, internal energy, and enthalpy holds irrespective of constant entropy. In (3.10.15) the proportionality to $e^{\frac{1}{2}}$ and $h^{\frac{1}{2}}$ requires the constant terms in e and h to be such that $e = 0$ and $h = 0$ when $T = 0$, as holds here by our choice $e = \frac{5}{2}RT$ and $h = \frac{7}{2}RT$.

3.11 Bibliographic Notes

Three established texts on thermodynamics are Pippard (1957), Sears & Salinger (1975), and Landau & Lifshitz (1980). Many of the works in Tables 13.3.1 and 13.4.1 have chapters or longer sections on thermodynamics (e.g., Rossini (1955), Liepmann & Roshko (1957), Zel'dovich & Raizer (1966, 1967), Thompson (1985), and Anderson (1989)). A text on physical gas dynamics is Vincenti & Kruger (1965), and there are several papers on physical gas dynamics in Lighthill (1997), volume 1, in the section "Gas dynamics interacting with gas physics." Two surveys of thermodynamics applied to fluid dynamics are Sullivan (1981) and Menikoff & Plohr (1989). A recent paper on high speed flow in which thermodynamic ideas play a large part is Mallinson, Gai & Mudford (1997).

4

Smooth Flow of an Ideal Fluid

4.1 Lattice of Special Cases

This chapter contains some general results about the smooth flow of an ideal fluid. Thus we consider regions in which the fluid's viscosity and thermal conductivity may be neglected and its density, pressure, velocity, etc. have no discontinuities or singularities in their values or derivatives, so that there are no vortex sheets or shocks. In later chapters we shall see many examples of such regions, which are typically bounded by shocks, boundary layers, and shear layers, and in most flows these regions are very extensive.

The equations of energy, mass, and momentum in (2.6.24) may be written

$$\frac{\partial s}{\partial t} + \mathbf{u} \cdot \nabla s = 0, \tag{4.1.1a}$$

$$\frac{\partial \rho}{\partial t} + \mathbf{u} \cdot \nabla \rho = -\rho \nabla \cdot \mathbf{u}, \tag{4.1.1b}$$

$$\frac{\partial \mathbf{u}}{\partial t} = \mathbf{u} \wedge \omega - \nabla \left(\frac{1}{2}\mathbf{u}^2\right) - \frac{1}{\rho}\nabla p + \mathbf{f}. \tag{4.1.1c}$$

Entropy is constant following each fluid particle, by (4.1.1a), but not necessarily throughout the fluid (i.e., the flow is isentropic but not necessarily homentropic). The flow is steady if the terms in $\partial/\partial t$ in (4.1.1) are identically zero, and irrotational if the vorticity $\omega = \nabla \wedge \mathbf{u}$ is identically zero; otherwise, the flow is unsteady or rotational.

We shall find that ω and ∇s are coupled in a simple way. Therefore we place in one class the flows with $\omega = 0$ and $\nabla s = 0$ (i.e., the irrotational homentropic flows). We consider flows of the types (i) $\partial/\partial t$ arbitrary; ω and ∇s arbitrary (i.e., general flow); (ii) $\partial/\partial t \equiv 0$; ω and ∇s arbitrary (i.e., steady flow); (iii) $\partial/\partial t$ arbitrary; $\omega = 0$, $\nabla s = 0$ (i.e., irrotational homentropic flow); and (iv) $\partial/\partial t \equiv 0$; $\omega = 0$, $\nabla s = 0$ (i.e., steady irrotational homentropic flow).

55

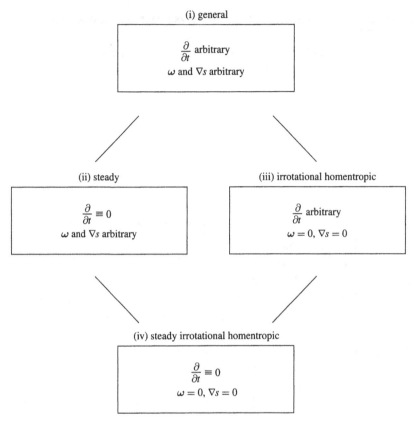

Fig. 4.1.1. Lattice of special cases of smooth flow of an ideal fluid. Bernoulli's equation takes the forms (4.3.3), (4.4.4), and (4.5.2) for cases (ii), (iii), and (iv), and there is no Bernoulli's equation for case (i), where it is replaced by Crocco's equation (4.2.9).

Thus (iv) is a special case of (ii) or (iii), and these are special cases of (i), as indicated by the lattice in Figure 4.1.1. We shall obtain general results for each of the four types of flow, notably Bernoulli's equation for types (ii), (iii), and (iv), for which it takes different forms, and Crocco's equation for type (i), for which there is no Bernoulli's equation. Physically, Bernoulli's equation corresponds to the existence of a conserved quantity, either on a streamline or particle path, or throughout the flow, and mathematically it corresponds to the existence of an integral of the governing differential equations. Thus we do not obtain an integral for general flow. Our terminology, though standard, is perhaps misleading, as there is no single Bernoulli's equation from which the various forms are obtainable as special cases. The results in this chapter hold for an arbitrary fluid, and we illustrate them by giving further details for a polytropic gas.

4.2 General Flow. Crocco's Equation

We begin by writing the momentum and mass equations in terms of s and the specific enthalpy h. Since $dh = T\,ds + \rho^{-1}\,dp$, we have $\nabla h = T\nabla s + \rho^{-1}\nabla p$, which from the momentum equation (4.1.1c) is

$$\nabla h = -\frac{\partial \mathbf{u}}{\partial t} + \mathbf{u} \wedge \boldsymbol{\omega} - \nabla\left(\frac{1}{2}\mathbf{u}^2\right) + T\nabla s + \mathbf{f}. \qquad (4.2.1)$$

Similarly, $Dh/Dt = T\,Ds/Dt + \rho^{-1}Dp/Dt$, and successive use of the energy equation (4.1.1a), the definition of the speed of sound c, and the mass equation (4.1.1b) gives $Dh/Dt = \rho^{-1}Dp/Dt = c^2\rho^{-1}D\rho/Dt = -c^2\nabla \cdot \mathbf{u}$. Thus

$$\frac{\partial h}{\partial t} = -\mathbf{u} \cdot \nabla h - c^2\nabla \cdot \mathbf{u}. \qquad (4.2.2)$$

Substituting (4.2.1) in (4.2.2), and making use of the identity $\mathbf{u} \cdot (\mathbf{u} \wedge \boldsymbol{\omega}) = 0$ and the energy equation in the form $\mathbf{u} \cdot \nabla s = -\partial s/\partial t$, we obtain

$$\frac{\partial h}{\partial t} = \frac{\partial}{\partial t}\left(\frac{1}{2}\mathbf{u}^2\right) + \mathbf{u} \cdot \nabla\left(\frac{1}{2}\mathbf{u}^2\right) - c^2\nabla \cdot \mathbf{u} + T\frac{\partial s}{\partial t} - \mathbf{u} \cdot \mathbf{f}. \qquad (4.2.3)$$

The time derivative of (4.2.1), minus the gradient of (4.2.3), gives

$$\nabla(c^2\nabla \cdot \mathbf{u}) - \frac{\partial^2 \mathbf{u}}{\partial t^2} = \nabla\left\{\frac{\partial}{\partial t}(\mathbf{u}^2) + \mathbf{u} \cdot \nabla\left(\frac{1}{2}\mathbf{u}^2\right) - \mathbf{u} \cdot \mathbf{f}\right\}$$
$$- \frac{\partial}{\partial t}(\mathbf{u} \wedge \boldsymbol{\omega} + \mathbf{f}) + \frac{\partial s}{\partial t}\nabla T - \frac{\partial T}{\partial t}\nabla s. \qquad (4.2.4)$$

This equation is a possible starting point for assumptions of steadiness, irrotationality, and homentropy, since certain terms then vanish. The left-hand side has the form of a wave operator acting on $\mathbf{u}$, and the right-hand side contains "source terms." Thus (4.2.4) is a form of wave analogy, or acoustic analogy. Its divergence is a scalar wave equation for $\nabla \cdot \mathbf{u}$. The curl of the momentum equation (4.1.1c) is

$$\frac{\partial \boldsymbol{\omega}}{\partial t} = \nabla \wedge (\mathbf{u} \wedge \boldsymbol{\omega}) + \nabla T \wedge \nabla s + \nabla \wedge \mathbf{f}. \qquad (4.2.5)$$

This is the vorticity equation, and the curl of (4.2.4) is its time derivative.

Assume now that the body force $\mathbf{f}$ is conservative and may be written in terms of a potential Φ as

$$\mathbf{f} = -\nabla\Phi. \qquad (4.2.6)$$

Substitution in (4.2.1) gives

$$\frac{\partial \mathbf{u}}{\partial t} = \mathbf{u} \wedge \omega - \nabla\left(h + \frac{1}{2}\mathbf{u}^2 + \Phi\right) + T\nabla s. \tag{4.2.7}$$

The quantity in parentheses is called the head, or total head, or total energy, and is denoted H. Thus

$$\begin{aligned} H &= h + \frac{1}{2}\mathbf{u}^2 + \Phi \\ &= e + \frac{p}{\rho} + \frac{1}{2}\mathbf{u}^2 + \Phi, \end{aligned} \tag{4.2.8}$$

and (4.2.7) is

$$\nabla H = -\frac{\partial \mathbf{u}}{\partial t} + \mathbf{u} \wedge \omega + T\nabla s. \tag{4.2.9}$$

This equation, or (4.2.7), is Crocco's equation. It gives

$$\begin{aligned} \mathbf{u} \cdot \nabla H &= -\frac{\partial}{\partial t}\left(\frac{1}{2}\mathbf{u}^2\right) + T\mathbf{u} \cdot \nabla s \\ &= -\frac{\partial}{\partial t}\left(\frac{1}{2}\mathbf{u}^2\right) - T\frac{\partial s}{\partial t}. \end{aligned} \tag{4.2.10}$$

Since (4.2.9) and (4.2.10) show that $\nabla H \neq 0$ and $\mathbf{u} \cdot \nabla H \neq 0$ in general, and similarly also $DH/Dt \neq 0$, it follows that, in general, H is not constant in a region or on a streamline or particle path. The most that can be said about H in complete generality is Crocco's equation.

4.3 Steady Flow

We now assume that the flow, though possibly rotational and nonhomentropic, is steady. Thus in the governing equations we replace $\partial/\partial t$ by 0 and D/Dt by $\mathbf{u} \cdot \nabla$, so that the energy equation becomes $\mathbf{u} \cdot \nabla s = 0$. Therefore on each streamline the entropy is constant. We still have $\mathbf{f} = -\nabla\Phi$ and $H = h + \frac{1}{2}\mathbf{u}^2 + \Phi$, and Crocco's equation (4.2.9) is now

$$\nabla H = \mathbf{u} \wedge \omega + T\nabla s. \tag{4.3.1}$$

Thus

$$\mathbf{u} \cdot \nabla H = 0. \tag{4.3.2}$$

Therefore on each streamline the head H is constant, a result we may write in the equivalent forms:

$$h + \frac{1}{2}\mathbf{u}^2 + \Phi = \text{const.},$$

$$e + \frac{p}{\rho} + \frac{1}{2}\mathbf{u}^2 + \Phi = \text{const.,} \qquad (4.3.3)$$

$$\int \frac{dp}{\rho} + \frac{1}{2}\mathbf{u}^2 + \Phi = \text{const.}$$

The last of these is a consequence of $dh = T ds + \rho^{-1} dp$ and $ds = 0$, whence $dh = \rho^{-1} dp$ and $h = \int \rho^{-1} dp$. Equation (4.3.3) on a streamline is Bernoulli's equation for steady flow of an ideal fluid. Since streamlines and particle paths are identical in steady flow, it states equally that H is constant following the fluid particles (i.e., that the flow is isoenergetic).

If H takes the same value on all streamlines (i.e., is constant in a region), the flow is said to be homenergetic. Steady flow of an ideal fluid is not in general homenergetic. However, if there is an upstream region of homenergetic flow, for example where the flow is rectilinear and at constant velocity and temperature, then all the streamlines from that region have the same value of H, and so the downstream flow is also homenergetic, as illustrated in Figure 4.3.1. Another type of steady flow that is homenergetic is steady irrotational homentropic flow, as we shall see in Section 4.5.

Equation (4.3.1) simplifies in a homenergetic region, because then $\nabla H = 0$, so that

$$\mathbf{u} \wedge \boldsymbol{\omega} + T\nabla s = 0. \qquad (4.3.4)$$

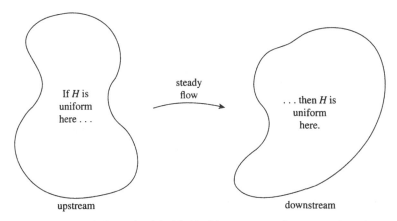

Fig. 4.3.1. Steady flow of an ideal fluid with an upstream homenergetic region.

Therefore downstream of a homenergetic region in a steady flow there is a simple coupling between entropy gradient and vorticity. For example, if $\nabla s \neq 0$ then $\omega \neq 0$; that is, imposition of an entropy gradient must produce vorticity. This is a useful point of view when considering the flow downstream of a shock of nonuniform strength. A flow in which $\omega \neq 0$ but $\mathbf{u} \wedge \omega = 0$ (i.e., in which ω is parallel to $\mathbf{u}$) is called a Beltrami flow. Such a flow, if steady and homenergetic, satisfies $\nabla s = 0$, by (4.3.4), and thus is homentropic. Beltrami flows are amenable to a mathematical theory of their streamline topology.

We now derive a single equation in $\mathbf{u}$, valid for all steady flows of an ideal fluid. The equations of energy, mass, and momentum are

$$\mathbf{u} \cdot \nabla s = 0, \tag{4.3.5a}$$

$$\mathbf{u} \cdot \nabla \rho + \rho \nabla \cdot \mathbf{u} = 0, \tag{4.3.5b}$$

$$\rho \mathbf{u} \cdot \nabla \mathbf{u} + \nabla p = \rho \mathbf{f}. \tag{4.3.5c}$$

The thermodynamic relation $dp = c^2 d\rho + (\partial p/\partial s)_\rho \, ds$ gives $\nabla p = c^2 \nabla \rho + (\partial p/\partial s)_\rho \nabla s$, which from (4.3.5a,b) gives

$$\mathbf{u} \cdot \nabla p = c^2 \mathbf{u} \cdot \nabla \rho = -\rho c^2 \nabla \cdot \mathbf{u}. \tag{4.3.6}$$

Therefore the scalar product of (4.3.5c) with $\mathbf{u}$ gives

$$c^2 \nabla \cdot \mathbf{u} = \mathbf{u} \cdot (\mathbf{u} \cdot \nabla \mathbf{u}) - \mathbf{u} \cdot \mathbf{f}. \tag{4.3.7}$$

The identity $\nabla(\tfrac{1}{2}\mathbf{u}^2) = \mathbf{u} \cdot \nabla \mathbf{u} + \mathbf{u} \wedge \omega$ gives

$$c^2 \nabla \cdot \mathbf{u} = \mathbf{u} \cdot \nabla\left(\tfrac{1}{2}\mathbf{u}^2\right) - \mathbf{u} \cdot \mathbf{f}. \tag{4.3.8}$$

In suffix notation, the last two equations are

$$c^2 \frac{\partial u_i}{\partial x_i} = u_i u_j \frac{\partial u_i}{\partial x_j} - u_i f_i. \tag{4.3.9}$$

The gradient of these equations is the steady version of (4.2.4), from which we could have begun.

Assume now that $\mathbf{f} = -\nabla \Phi$. On each streamline we may regard h as a function of c, since the flow is isentropic. Hence Bernoulli's equation (4.3.3a) gives c as a function of Φ and $\mathbf{u}^2$. We shall label the streamlines by a symbol ψ, which in three-dimensional flow represents two parameters and in two-dimensional or axisymmetric flow represents one parameter, which could be the stream function. Bernoulli's equation therefore gives the functional relation

$$c^2 = c^2(\psi, \Phi, \mathbf{u}^2). \tag{4.3.10}$$

Thus (4.3.8) is

$$c^2(\psi, \Phi, \mathbf{u}^2)\nabla \cdot \mathbf{u} = \mathbf{u} \cdot \nabla\left(\tfrac{1}{2}\mathbf{u}^2 + \Phi\right). \qquad (4.3.11)$$

This equation is most useful when $\mathbf{u}$ can be expressed in terms of a single scalar function (e.g., a stream function or potential). In conjunction with boundary conditions and jump conditions, it provides a starting point for solving boundary value problems, for example in linear or nonlinear aerofoil theory.

To proceed further, we need the properties of the fluid. For a polytropic gas with ratio of specific heats γ, we have $h = c^2/(\gamma - 1)$, so that Bernoulli's equation is

$$\frac{c^2}{\gamma - 1} + \frac{1}{2}\mathbf{u}^2 + \Phi = \text{const.} \qquad (4.3.12)$$

Let us ignore body forces (i.e., put $\Phi = 0$). The constant in (4.3.12) may be specified by giving the value of c at a hypothetical point where $\mathbf{u} = 0$, that is, at the stagnation condition. Thus the stagnation sound speed, $c_0(\psi)$, which may vary from one streamline to another, is defined by

$$\frac{c^2}{\gamma - 1} + \frac{1}{2}\mathbf{u}^2 = \frac{c_0^2(\psi)}{\gamma - 1}. \qquad (4.3.13)$$

Therefore

$$c^2 = c_0^2(\psi) - \tfrac{1}{2}(\gamma - 1)\mathbf{u}^2. \qquad (4.3.14)$$

With $\Phi = 0$ and c given by (4.3.14), Equation (4.3.11) becomes

$$\left\{c_0^2(\psi) - \tfrac{1}{2}(\gamma - 1)\mathbf{u}^2\right\}\nabla \cdot \mathbf{u} = \mathbf{u} \cdot \nabla\left(\tfrac{1}{2}\mathbf{u}^2\right). \qquad (4.3.15)$$

This equation, cubic in the components of $\mathbf{u}$ and their spatial derivatives, is a basic equation satisfied by the velocity field $\mathbf{u}$ in steady flow of an ideal polytropic gas.

4.4 Irrotational Homentropic Flow

For flow that is irrotational and homentropic, but possibly unsteady, we put $\omega = 0$ and $\nabla s = 0$ in (4.1.1), so that the energy equation becomes $\partial s/\partial t = 0$ and s takes a single value for all time in the fluid. We still assume that $\mathbf{f} = -\nabla\Phi$, whence $H = h + \tfrac{1}{2}\mathbf{u}^2 + \Phi$. Crocco's equation (4.2.9) loses the terms $\mathbf{u} \wedge \omega$ and $T\nabla s$ to become

$$\nabla H = -\frac{\partial \mathbf{u}}{\partial t}. \qquad (4.4.1)$$

Since the flow is irrotational, the velocity field is the gradient of a velocity potential $\phi(\mathbf{x}, t)$, possibly multiple-valued. Thus

$$\mathbf{u} = \nabla\phi. \tag{4.4.2}$$

The function ϕ is indeterminate to the extent that an arbitrary function of time may be added to it without affecting $\mathbf{u}$. Equation (4.4.1) gives

$$\nabla\left(\frac{\partial\phi}{\partial t} + H\right) = 0. \tag{4.4.3}$$

Therefore $\partial\phi/\partial t + H$ is constant throughout the fluid. Since s is constant, the flow is barotropic (i.e., $\rho = \rho(p)$ and $h = h(p) = \int \rho^{-1}\, dp$), and we also have $\mathbf{u}^2 = |\nabla\phi|^2$. Hence (4.4.3) gives

$$\frac{\partial\phi}{\partial t} + \frac{1}{2}|\nabla\phi|^2 + \int \frac{dp}{\rho} + \Phi = \text{const.} \tag{4.4.4}$$

This is Bernoulli's equation for irrotational homentropic flow of an ideal fluid. The constant on the right has the same value throughout the fluid, unlike the constant in (4.3.3). It may contain an arbitrary function of time, but this may be incorporated into $\partial\phi/\partial t$, and hence into ϕ, without affecting $\mathbf{u}$.

To derive a single equation in ϕ, suitable for boundary-value problems, we may start with the mass equation in the form $c^2\nabla \cdot \mathbf{u} = -Dh/Dt$ and Bernoulli's equation (4.4.4) in the form $\partial\phi/\partial t + \frac{1}{2}|\nabla\phi|^2 + h + \Phi = h_0$, where the constant on the right has been written h_0. Elimination of h gives

$$c^2\nabla^2\phi = \left(\frac{\partial}{\partial t} + (\nabla\phi) \cdot \nabla\right)\left(\frac{\partial\phi}{\partial t} + \frac{1}{2}|\nabla\phi|^2 + \Phi\right), \tag{4.4.5}$$

that is,

$$c^2\nabla^2\phi - \frac{\partial^2\phi}{\partial t^2} = \nabla\phi \cdot \nabla\left(\frac{1}{2}|\nabla\phi|^2\right) + \frac{\partial}{\partial t}\left(|\nabla\phi|^2\right) + \left(\frac{\partial}{\partial t} + (\nabla\phi) \cdot \nabla\right)\Phi, \tag{4.4.6}$$

or

$$\left(c^2\delta_{ij} - \frac{\partial\phi}{\partial x_i}\frac{\partial\phi}{\partial x_j}\right)\frac{\partial^2\phi}{\partial x_i\partial x_j} - 2\frac{\partial\phi}{\partial x_i}\frac{\partial^2\phi}{\partial t\partial x_i} - \frac{\partial^2\phi}{\partial t^2} = \left(\frac{\partial}{\partial t} + \frac{\partial\phi}{\partial x_i}\frac{\partial}{\partial x_i}\right)\Phi, \tag{4.4.7}$$

or

$$c^2\nabla \cdot \mathbf{u} - \frac{\partial^2\phi}{\partial t^2} = \mathbf{u} \cdot (\mathbf{u} \cdot \nabla\mathbf{u}) + \frac{\partial}{\partial t}(\mathbf{u}^2) - \mathbf{u} \cdot \mathbf{f} + \frac{\partial\Phi}{\partial t}. \tag{4.4.8}$$

Alternatively, we may obtain these equations from (4.2.4). With the unsteady terms omitted, (4.4.8) becomes (4.3.7).

For a polytropic gas with ratio of specific heats γ, Bernoulli's equation (4.4.4) is

$$\frac{\partial \phi}{\partial t} + \frac{1}{2}|\nabla \phi|^2 + \frac{c^2}{\gamma - 1} + \Phi = \text{const.} \qquad (4.4.9)$$

If there are no body forces, and the stagnation sound speed is c_0, this is

$$\frac{\partial \phi}{\partial t} + \frac{1}{2}|\nabla \phi|^2 + \frac{c^2}{\gamma - 1} = \frac{c_0^2}{\gamma - 1}, \qquad (4.4.10)$$

which can be rearranged to give

$$c^2 = c_0^2 - (\gamma - 1)\left(\frac{\partial \phi}{\partial t} + \frac{1}{2}|\nabla \phi|^2\right). \qquad (4.4.11)$$

Here c_0 is a constant for the whole fluid. Equation (4.4.6) or (4.4.7), with $\Phi = 0$ and c^2 given by (4.4.11), is a basic equation for irrotational homentropic flow of an ideal polytropic gas.

4.5 Steady Irrotational Homentropic Flow

When the flow is steady, irrotational, and homentropic we combine the results in the last two sections. Thus the velocity is $\mathbf{u} = \nabla \phi$, the body force per unit mass is $\mathbf{f} = -\nabla \Phi$, the head is $H = h + \frac{1}{2}\mathbf{u}^2 + \Phi$, and Crocco's equation (4.2.9) loses all the terms on the right to become simply

$$\nabla H = 0. \qquad (4.5.1)$$

Therefore H is constant throughout the fluid (i.e., the flow is homenergetic). For later reference, let us write out Equations (4.4.4)–(4.4.11) with the terms in $\partial/\partial t$ omitted. Bernoulli's equation (4.4.4) is

$$\frac{1}{2}|\nabla \phi|^2 + \int \frac{dp}{\rho} + \Phi = \text{const.}, \qquad (4.5.2)$$

that is,

$$h = h_0 - \frac{1}{2}|\nabla \phi|^2 - \Phi. \qquad (4.5.3)$$

The single equation for ϕ is

$$c^2 \nabla^2 \phi = \nabla \phi \cdot \nabla\left(\frac{1}{2}|\nabla \phi|^2\right) + \nabla \phi \cdot \nabla \Phi, \qquad (4.5.4)$$

that is,

$$\left(c^2 \delta_{ij} - \frac{\partial \phi}{\partial x_i} \frac{\partial \phi}{\partial x_j} \right) \frac{\partial^2 \phi}{\partial x_i \partial x_j} = \frac{\partial \phi}{\partial x_i} \frac{\partial \Phi}{\partial x_i}. \qquad (4.5.5)$$

This is simply (4.3.8) or (4.3.9) with $\mathbf{u} = \nabla \phi$. In two-dimensional flow, the equations $\nabla \wedge \mathbf{u} = 0$ and (4.5.5) may be written in the variables (x, y) and (u, v), with suffices for partial derivatives, as

$$u_y - v_x = 0,$$
$$(c^2 - u^2)u_x - uv(u_y + v_x) + (c^2 - v^2)v_y = -\mathbf{u} \cdot \mathbf{f}. \qquad (4.5.6)$$

For a polytropic gas, with $h = c^2/(\gamma - 1)$, Bernoulli's equation (4.5.2) is

$$\frac{1}{2}|\nabla \phi|^2 + \frac{c^2}{\gamma - 1} + \Phi = \text{const.} \qquad (4.5.7)$$

In the absence of body forces (i.e., with $\Phi = 0$), we use the stagnation sound speed c_0 to obtain

$$\frac{1}{2}|\nabla \phi|^2 + \frac{c^2}{\gamma - 1} = \frac{c_0^2}{\gamma - 1}, \qquad (4.5.8)$$

that is,

$$c^2 = c_0^2 - \frac{1}{2}(\gamma - 1)|\nabla \phi|^2. \qquad (4.5.9)$$

Substitution of (4.5.9) into (4.5.4) or (4.5.5), with no body forces, gives a basic equation for steady irrotational homentropic flow of an ideal polytropic gas.

In later chapters we make frequent use of the equations in this section, for example in our treatment of flow past an aerofoil or round a bend and in the theory of characteristics for steady two-dimensional flow.

4.6 Bibliographic Notes

General consequences of the equations governing smooth flow are obtained in many of the works in Tables 13.3.1 and 13.4.1 up to about 1960, for example Ward (1955), Tsien (1958), Miles (1959), Serrin (1959), and Schiffer (1960). See also, for example, the chapter "High speed flow of air" in Paterson (1983).

4.7 Further Results and Exercises

1. Obtain Crocco's equation for the steady motion of an inviscid fluid under conservative forces, that is,

$$\mathbf{u} \wedge \omega = \nabla H - T\nabla s.$$

In a steady flow of a perfect gas with constant total energy, the streamlines are given by constant values of the function $\psi(\mathbf{x})$. Show that

$$\mathbf{u} \wedge \boldsymbol{\omega} = -T \frac{ds}{d\psi} \nabla \psi.$$

Deduce that, for two-dimensional flow, the vorticity is

$$\omega = \frac{T}{|\mathbf{u}|^2} \frac{ds}{d\psi} (\mathbf{u} \wedge \nabla \psi).$$

Let the density of the gas be ρ, and let a reference density be ρ_0. By considering conservation of mass, show that

$$|\omega| = \frac{\rho T}{\rho_0} \left| \frac{ds}{d\psi} \right|.$$

Deduce that on any streamline the vorticity is proportional to the pressure.

2. Show that, for a perfect gas with constant specific heats, $p \propto \rho^\gamma \exp(s/c_v)$. Write down the equations of motion for unsteady one-dimensional fluid flow, not assuming constant entropy, and prove that, in new independent variables time t and entropy s, the equations are

$$\rho_t x_s + \rho u_s = 0, \qquad \rho u_t x_s + p_s = 0, \qquad x_t - u = 0.$$

Deduce that ρx_s is independent of time.

3. Derive an expression for the rate of change of circulation around a closed circuit in an ideal compressible fluid when temperature and entropy gradients are present.

4. Assuming Bernoulli's equation and the equations of motion and continuity for steady irrotational two-dimensional flow of a compressible fluid, show that in polar coordinates (r, θ) the equation for the potential $\phi(r, \theta)$ is

$$\left(c^2 - \phi_r^2 \right) \phi_{rr} + \left(c^2 - r^{-2}\phi_\theta^2 \right) r^{-2} \phi_{\theta\theta}$$
$$- 2r^{-2} \phi_r \phi_\theta \phi_{r\theta} + \left(c^2 + r^{-2}\phi_\theta^2 \right) r^{-1} \phi_r = 0.$$

Show that two types of solution exist of the form $\phi = r f(\theta)$ and that these solutions may be used to describe flow in contact with a rigid boundary having a sharp convex corner. Investigate these solutions as far as you can.

5. Starting from the laws of thermodynamics and the vector equations governing the steady flow of an inviscid perfect gas, prove that

$$\mathbf{u} \wedge \boldsymbol{\omega} = \nabla \left(e + \frac{p}{\rho} + \frac{1}{2}\mathbf{u}^2 \right) - T\nabla s.$$

Suppose that $e + p/\rho + \frac{1}{2}\mathbf{u}^2$ is everywhere constant and the motion is two-dimensional. Show that s is constant along streamlines. Let the stream function be ψ. Show that $|\omega|/(pds/d\psi)$ is everywhere constant. If the maximum possible speed of the gas is $q_{\max}$, and the ratio of specific heats is γ, deduce that, for some function $f(\psi)$, the flow satisfies the equation

$$|\omega| = f(\psi)\big(q_{\max}^2 - \mathbf{u}^2\big)^{\gamma/(\gamma-1)}.$$

5

Characteristic Surfaces and Rays

5.1 Nonsmooth Flow

The flow at a point is nonsmooth if at least one fluid-dynamical or thermo-dynamic variable has there a discontinuity or singularity in its value or in a derivative of some order with respect to space or time. Points of nonsmoothness may be isolated, or they may form lines or surfaces. The following discussion concentrates on nonsmoothness across surfaces. We shall be particularly concerned with surfaces across which the pressure has a discontinuity in its value or in a derivative. There will be corresponding discontinuities in the other variables (i.e., velocity, density, temperature, etc.).

One type of surface of possible nonsmoothness is a characteristic surface of a set of hyperbolic partial differential equations and is determined by the differential equations themselves. We shall give a precise definition of a characteristic surface in the next section. Nonlinear hyperbolic equations often do not allow a discontinuity in a dependent variable on a characteristic surface, but only in the derivatives of the variable. For example, the differential equations for the flow of an ideal fluid are nonlinear and hyperbolic. They allow a discontinuity in the derivatives of the pressure on a characteristic surface but do not allow a discontinuity in the pressure itself. A discontinuity in a derivative of a variable, but not in the variable itself, is called a weak discontinuity. In compressible flow, the most frequently encountered weak discontinuities occur on surfaces that propagate relative to the fluid at the local speed of sound and are called Mach waves. Thus discontinuities in the derivatives of pressure, velocity, density, temperature, etc. are usually Mach waves.

A second type of surface of nonsmoothness is one that is not determined by the differential equations being used, but by jump conditions, often obtained, as in Chapter 2, from conservation laws. For example, the equations governing the flow of an ideal fluid cannot represent the internal structure of a shock, which is determined by thermoviscous and relaxation effects. Since a shock is usually thin, it may usually be represented by a surface separating two regions of smooth flow of an ideal fluid, provided that, at this surface, the jump conditions for a shock are satisfied.

Shocks and characteristic surfaces do not in general coincide, but they do so in the limit of zero shock strength, when the shock wave becomes a sound wave, propagating at the local speed of sound. Thus the distinction between shocks and characteristic surfaces, which arises in the full nonlinear equations, disappears in the linearised acoustic equations describing small perturbations to a flow. Therefore in linear acoustics the terms Mach wave and shock wave may be used interchangeably, and jumps in pressure and velocity, not just in their derivatives, may occur on characteristic surfaces. The jumps are still called weak, because the equations have been linearised. A full strength shock (i.e., a shock that is not weak and satisfies the full nonlinear equations, supplemented by jump conditions) cannot occur on a characteristic surface.

We present in this chapter the theory of characteristic surfaces, and in the next the theory of shocks. We distinguish between (a) problems in two independent variables, that is, two space coordinates, or one space coordinate and time, and (b) problems in more than two independent variables, for example three space coordinates, or two or three space coordinates and time. Since characteristic surfaces have dimension one less than the number of independent variables (i.e., they are hypersurfaces), it follows that in case (a) the characteristics are curves. The equations then reduce to ordinary differential equations on these curves. When there are as many families of characteristics as the number of dependent variables, the ordinary differential equations along the different characteristics are sufficient in number to determine the solution of the original partial differential equations, which are then said to be totally hyperbolic. This approach is the Friedrichs theory for a totally hyperbolic system. In case (b), the characteristics are surfaces of dimension at least two, and the problem cannot be reduced to a set of ordinary differential equations. This chapter contains the theory for any number of independent variables, and Chapter 10 contains the Friedrichs theory, for two independent variables.

5.2 Characteristic Surfaces

In this section we determine the characteristic surfaces for an ideal fluid in arbitrary motion. Thus the flow may be unsteady, three-dimensional, rotational, and nonhomentropic. The dependent variables, taken to be density ρ, three components of velocity u, v, w, and specific entropy s, will be combined into a single vector (ρ, u, v, w, s) and written $(u_0, u_1, u_2, u_3, u_4)$ or $\mathbf{u}$. The velocity vector (u, v, w) (i.e., (u_1, u_2, u_3)) is written $\mathbf{u}'$. Thus

$$\mathbf{u} = (u_0, u_1, u_2, u_3, u_4) = (\rho, u, v, w, s),$$
$$\mathbf{u}' = (u_1, u_2, u_3) = (u, v, w).$$
(5.2.1)

The speed of sound c is defined by $c^2 = (\partial p / \partial \rho)_s$, and we define a and b by

$$\frac{\nabla p}{\rho} = a \nabla \rho + b \nabla s,$$
(5.2.2)

that is,

$$a = \frac{c^2}{\rho}, \qquad b = \frac{1}{\rho}\left(\frac{\partial p}{\partial s}\right)_\rho.$$
(5.2.3)

We ignore body forces, as they do not affect the theory of characteristic surfaces. The equations of conservation of mass, momentum, and energy for an ideal fluid, as given in (2.6.24), are then

$$\frac{\partial \rho}{\partial t} + \mathbf{u}' \cdot \nabla \rho + \rho \nabla \cdot \mathbf{u}' = 0,$$

$$\frac{\partial \mathbf{u}'}{\partial t} + \mathbf{u}' \cdot \nabla \mathbf{u}' + a \nabla \rho + b \nabla s = 0,$$
(5.2.4)

$$\frac{\partial s}{\partial t} + \mathbf{u}' \cdot \nabla s = 0.$$

In matrix form, with the identity matrix written I and with partial differentiation denoted with subscripts t, x, y, z, these equations are

$$
I \begin{pmatrix} \rho \\ u \\ v \\ w \\ s \end{pmatrix}_t
+ \begin{pmatrix} u & \rho & & & \\ a & u & & & b \\ & & u & & \\ & & & u & \\ & & & & u \end{pmatrix} \begin{pmatrix} \rho \\ u \\ v \\ w \\ s \end{pmatrix}_x
+ \begin{pmatrix} v & & \rho & & \\ & v & & & \\ a & & v & & b \\ & & & v & \\ & & & & v \end{pmatrix} \begin{pmatrix} \rho \\ u \\ v \\ w \\ s \end{pmatrix}_y
$$

$$
+ \begin{pmatrix} w & & & \rho & \\ & w & & & \\ & & w & & \\ a & & & w & b \\ & & & & w \end{pmatrix} \begin{pmatrix} \rho \\ u \\ v \\ w \\ s \end{pmatrix}_z
= \begin{pmatrix} 0 \\ 0 \\ 0 \\ 0 \\ 0 \end{pmatrix}.
$$
(5.2.5)

The four matrices here, each of size 5×5, will be written A^0, A^1, A^2, and A^3. The superscripts do not denote powers, although $A^0 = I$. With $\mathbf{u} = (\rho, u, v, w, s)$ as a column vector, (5.2.5) is

$$A^0 \mathbf{u}_t + A^1 \mathbf{u}_x + A^2 \mathbf{u}_y + A^3 \mathbf{u}_z = 0. \tag{5.2.6}$$

The independent variables t, x, y, and z will be combined into a single vector (t, x, y, z) and written (x_0, x_1, x_2, x_3) or $\mathbf{x}$, and the components of the matrices A^k will be written (a_{ij}^k), with k taking the values 0, 1, 2, 3 and i and j the values 0, 1, 2, 3, 4. In what follows, i, j, and k always take these values, and where appropriate we use the summation convention for repeated indices. Thus

$$\begin{aligned} \mathbf{x} = (x_k) = (x_0, x_1, x_2, x_3) = (t, x, y, z), \\ A^k = \left(a_{ij}^k \right), \end{aligned} \tag{5.2.7}$$

and (5.2.6) is

$$A^k \frac{\partial \mathbf{u}}{\partial x_k} = 0, \tag{5.2.8}$$

that is,

$$a_{ij}^k \frac{\partial u_j}{\partial x_k} = 0. \tag{5.2.9}$$

Across a characteristic surface, $\mathbf{u}$ need not be smooth even though (5.2.9) is satisfied on each side. Let the equation of a characteristic surface be $\phi(\mathbf{x}) = 0$. Recall that $\mathbf{x}$ is an arbitrary point in a four-dimensional space and with each $\mathbf{x}$ is associated a five-dimensional vector $\mathbf{u} = \mathbf{u}(\mathbf{x})$. The surface $\phi(\mathbf{x}) = 0$ occupies a three-dimensional subspace of four-dimensional $\mathbf{x}$-space. The normal $\mathbf{n} = (n_k) = (n_0, n_1, n_2, n_3)$ to the surface is parallel to the gradient of $\phi(\mathbf{x})$ (i.e., $n_k \propto \partial \phi / \partial x_k$). We now determine an equation satisfied by those $\mathbf{n}$ for which the function $\mathbf{u}(\mathbf{x})$ need not be smooth across the surface $\phi(\mathbf{x}) = 0$.

One method of finding the equation for $\mathbf{n}$ makes use of linear combinations. In (5.2.9), we multiply the i-equation by a coefficient l_i, which at this stage is arbitrary, and sum over i. Equivalently, we take the scalar product of the equations with an arbitrary vector $\mathbf{l} = (l_i)$. The result is the single equation

$$l_i a_{ij}^k \frac{\partial u_j}{\partial x_k} = 0. \tag{5.2.10}$$

Given a direction $\mathbf{m} = (m_k)$, the differential operator $m_k \partial / \partial x_k$ is the directional derivative in the direction $\mathbf{m}$. Let us concentrate on u_j for a particular value of j. Since u_j is acted on in (5.2.10) by the differential operator $l_i a_{ij}^k \partial / \partial x_k$, the only aspect of u_j that enters (5.2.10) is the derivative in the direction

$$\left(l_i a_{ij}^0, l_i a_{ij}^1, l_i a_{ij}^2, l_i a_{ij}^3\right). \tag{5.2.11}$$

Now consider all five components u_j. Each j has an associated direction (5.2.11), that is, in total there are five directions, and we ask whether they can all be perpendicular to a single direction $\mathbf{n} = (n_k)$. Since the condition for the vectors $\mathbf{m} = (m_k)$ and $\mathbf{n} = (n_k)$ to be perpendicular is $m_k n_k = 0$, it follows that the five directions are all perpendicular to $\mathbf{n}$ if, for all j,

$$l_i a_{ij}^k n_k = 0. \tag{5.2.12}$$

So far the components l_i of the vector $\mathbf{l}$ are arbitrary, except that, to correspond to a direction, they must not all be zero. Therefore we are asking whether there exist coefficients l_i, not all zero, such that (5.2.12) holds for all j. The question is answered by regarding $(a_{ij}^k n_k)$ as a 5×5 matrix in which i labels the rows, j labels the columns, and the (i, j) element is a sum of four terms, that is, $a_{ij}^0 n_0 + a_{ij}^1 n_1 + a_{ij}^2 n_2 + a_{ij}^3 n_3$. The required coefficients l_i exist if the determinant of the 5×5 matrix is zero:

$$\det\left(a_{ij}^k n_k\right) = 0. \tag{5.2.13}$$

When $\mathbf{n}$ satisfies (5.2.13), there is a linear combination of Equations (5.2.9) that contains only directional derivatives tangential to a surface that is everywhere perpendicular to $\mathbf{n}$. The resulting equation places no restriction on derivatives of $\mathbf{u}$ normal to this surface (i.e., tangential to $\mathbf{n}$). Hence outward derivatives may be discontinuous across the surface (i.e., they may have jumps). The vector $\mathbf{n}$ was defined to be perpendicular to the surface $\phi(\mathbf{x}) = 0$, so that (5.2.13) is equivalent to

$$\det\left(a_{ij}^k \frac{\partial \phi}{\partial x_k}\right) = 0. \tag{5.2.14}$$

Thus (5.2.14) is the condition for the surface $\phi(\mathbf{x}) = 0$ to be a characteristic surface for the equations $a_{ij}^k \partial u_j / \partial x_k = 0$.

Another method of finding the equation for $\mathbf{n}$ involves a change of variables. Consider a smooth change of variables from $\mathbf{x}$ to $\mathbf{x}'$, that is, from (x_k) to (x_k'). To a given value of x_0' there corresponds a surface in $\mathbf{x}$-space consisting of those $\mathbf{x}$ such that the given value of x_0' is taken by the function $x_0'(\mathbf{x})$. The derivative of $\mathbf{u}(\mathbf{x})$ in a direction perpendicular to this surface in $\mathbf{x}$-space may be obtained by expressing $\mathbf{u}$ in terms of $\mathbf{x}'$ and calculating $\partial \mathbf{u}/\partial x_0'$. We ask whether the value of $\partial \mathbf{u}/\partial x_0'$ is determined by the partial differential equations $a_{ij}^k \partial u_j / \partial x_k = 0$ (i.e., by (5.2.9)). On making the above change of variables, and using the chain rule, we obtain

$$a_{ij}^k \frac{\partial x_m'}{\partial x_k} \frac{\partial u_j}{\partial x_m'} = 0. \tag{5.2.15}$$

We collect together on the left the terms in $\partial u_j/\partial x_0'$ for all j, and on the right the remaining terms, which we denote by the symbol $\{L_{ij}(\partial/\partial x_1', \partial/\partial x_2', \partial/\partial x_3')\}u_j$, where i is a free suffix and j is summed. Then (5.2.15) is

$$a_{ij}^k \frac{\partial x_0'}{\partial x_k} \frac{\partial u_j}{\partial x_0'} = \left\{ L_{ij}\left(\frac{\partial}{\partial x_1'}, \frac{\partial}{\partial x_2'}, \frac{\partial}{\partial x_3'} \right) \right\} u_j. \tag{5.2.16}$$

The normals in x-space to surfaces corresponding to a fixed value of x_0' will be denoted by $\mathbf{n}$, so that we may take $n_k = \partial x_0'/\partial x_k$. Then (5.2.16) is

$$a_{ij}^k n_k \frac{\partial u_j}{\partial x_0'} = \left\{ L_{ij}\left(\frac{\partial}{\partial x_1'}, \frac{\partial}{\partial x_2'}, \frac{\partial}{\partial x_3'} \right) \right\} u_j. \tag{5.2.17}$$

We regard (5.2.17) as an equation for the column vector $\partial \mathbf{u}/\partial x_0'$, which occurs multiplied by the 5×5 matrix $(a_{ij}^k n_k)$, where i labels the rows and j the columns. The equation does not determine the vector $\partial \mathbf{u}/\partial x_0'$ uniquely when the determinant of the matrix is zero (i.e., when $\det(a_{ij}^k n_k) = 0$), or, equivalently, when an eigenvalue of the matrix is zero. In this case the value of $\partial \mathbf{u}/\partial x_0'$ is not determined by the values of $\mathbf{u}$ on a surface of constant x_0' but may contain an arbitrary multiple of an eigenvector of the matrix corresponding to the eigenvalue zero; that is, it may contain a null vector of the singular matrix $(a_{ij}^k n_k)$. Hence the equation $\det(a_{ij}^k n_k) = 0$ is the condition for $\mathbf{n}$ to be perpendicular to a characteristic surface, as we found in (5.2.13). A characteristic surface therefore consists of those $\mathbf{x}$ for which $x_0'(\mathbf{x})$ takes a given value when the normal $\mathbf{n}$ defined by $n_k = \partial x_0'/\partial x_k$ satisfies $\det(a_{ij}^k n_k) = 0$. Thus we need only change the notation from x_0' to ϕ to obtain the earlier result (5.2.14) that a surface of constant $\phi(\mathbf{x})$ is a characteristic surface if $\det(a_{ij}^k \partial\phi/\partial x_k) = 0$. Note that if $\det(a_{ij}^k n_k) = 0$, then some nontrivial linear combination of the rows of the matrix $(a_{ij}^k n_k)$ is zero, that is, for some values of l_i, not all zero, we have $l_i a_{ij}^k n_k = 0$. Thus we are back to the starting point of the method of linear combinations, and the two methods are more similar than they appear at first.

It is convenient to distinguish four types of characteristic surface, of which two are infinitesimal (i.e., local). These two determine possible tangent directions to noninfinitesimal (i.e., global) characteristic surfaces. At any point (t, x, y, z) we take any value of $\mathbf{n}$ that at the point satisfies $\det(a_{ij}^k n_k) = 0$, and we construct the first type of characteristic surface as an infinitesimal planar element perpendicular to $\mathbf{n}$, as shown in Figure 5.2.1a. The vertical axis in the figure represents t (i.e., x_0) and the horizontal axes represent x, y, and z (i.e.,

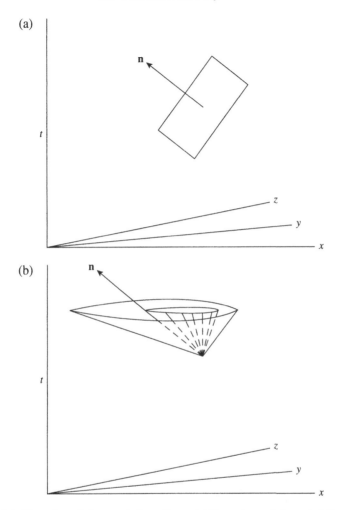

Fig. 5.2.1. Normals and characteristic surfaces. (a) Normal **n** and characteristic planar element. (b) Cone of normals and Monge cone (i.e., envelope of planar elements perpendicular to the normals). (c) Ray conoid and a Monge cone tangent to it. (d) Arbitrary characteristic surface and a Monge cone tangent to it.

x_1, x_2, and x_3). Such a surface describes locally planar propagation in an infinitesimal region and may be called a characteristic planar element. It provides a local approximation to a smooth part of a typical characteristic surface, and if extended to a whole plane it becomes a tangent plane to that surface.

The second type of infinitesimal characteristic surface is constructed from all the values of **n** that at a point satisfy $\det(a_{ij}^k n_k) = 0$. These **n** form the cone of normals at the point, and the corresponding characteristic planar elements, each perpendicular to such an **n**, have as their envelope another cone, different

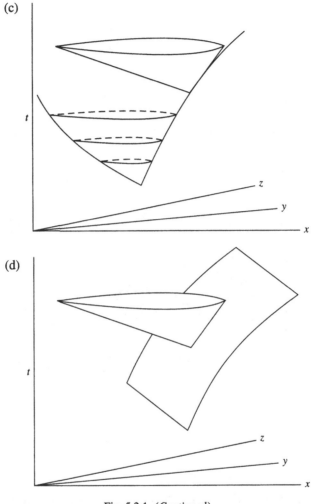

Fig. 5.2.1. (*Continued*)

from the cone of normals and of infinitesimal size, called the Monge cone or ray cone, shown in Figure 5.2.1b. This is the second type of characteristic surface, describing propagation in an infinitesimal region away from an instantaneous point disturbance. An arbitrary characteristic surface is tangential at each point to the Monge cone at that point.

The third type of characteristic surface is obtained from the Monge cone at a point by extending it to a global conelike characteristic surface, called a ray conoid, as shown in Figure 5.2.1c. Thus a ray conoid is tangential at each point to the Monge cone at that point, and the tangent directions away from its vertex are those of the Monge cone at the vertex. A ray conoid describes global propagation away from an instantaneous point disturbance.

Finally, we show in Figure 5.2.1d an arbitrary characteristic surface, constructed either from Monge cones tangent to it or from characteristic planar elements. The former construction is one version of the Huygens construction of a wavefront from a succession of point disturbances, and the latter is equivalent to specifying the surface as the envelope of its tangent planes.

Let us evaluate $\det(a_{ij}^k n_k)$ for the fluid-dynamical equations (5.2.4)–(5.2.5). Define m by

$$m = n_0 + u n_1 + v n_2 + w n_3. \qquad (5.2.18)$$

Then

$$
\begin{aligned}
\det(a_{ij}^k n_k) &= \det(A^k n_k) \\
&= \det(A^0 n_0 + A^1 n_1 + A^2 n_2 + A^3 n_3) \\
&= \begin{vmatrix}
m & \rho n_1 & \rho n_2 & \rho n_3 & 0 \\
a n_1 & m & 0 & 0 & b n_1 \\
a n_2 & 0 & m & 0 & b n_2 \\
a n_3 & 0 & 0 & m & b n_3 \\
0 & 0 & 0 & 0 & m
\end{vmatrix} \\
&= m^3 \big(m^2 - c^2 (n_1^2 + n_2^2 + n_3^2) \big).
\end{aligned}
\qquad (5.2.19)
$$

Therefore $\det(a_{ij}^k n_k) = 0$ when $\mathbf{n}$ satisfies either $m = 0$ or $m^2 = c^2(n_1^2+n_2^2+n_3^2)$, that is,

$$
\begin{aligned}
n_0 + u n_1 + v n_2 + w n_3 &= 0, \\
(n_0 + u n_1 + v n_2 + w n_3)^2 &= c^2 (n_1^2 + n_2^2 + n_3^2).
\end{aligned}
\qquad (5.2.20)
$$

Thus the cone of normals at (t, x, y, z) consists of all $\mathbf{n}$ satisfying either of the Equations (5.2.20).

5.3 The Monge Cone

We defined the Monge cone as the envelope of infinitesimal planar elements perpendicular to directions $\mathbf{n}$ on the cone of normals. We now show that if the equation of the cone of normals is $F(\mathbf{n}) = 0$, then the Monge cone is generated by the directions $\partial F / \partial \mathbf{n}$, that is, by $(\partial F/\partial n_0, \partial F/\partial n_1, \partial F/\partial n_2, \partial F/\partial n_3)$, also written $(\partial F/\partial n_k)$ or $\nabla_{\mathbf{n}} F$.

Note first that $F(\mathbf{n}) = \det(a_{ij}^{k} n_k)$, so that $F(\mathbf{n})$ is a homogeneous function of $\mathbf{n}$, and therefore, by Euler's theorem for homogeneous functions, $\mathbf{n} \cdot \partial F/\partial \mathbf{n} \propto F(\mathbf{n}) = 0$ (i.e., $\partial F/\partial \mathbf{n}$ is perpendicular to $\mathbf{n}$). Moreover, $\partial F/\partial \mathbf{n}$ is perpendicular to all neighbouring $\mathbf{n}$ on the cone of normals, because the condition $F(\mathbf{n} + d\mathbf{n}) = 0$ gives $F(\mathbf{n}) + d\mathbf{n} \cdot \partial F/\partial \mathbf{n} = 0$, and hence $d\mathbf{n} \cdot \partial F/\partial \mathbf{n} = 0$, or $(\mathbf{n} + d\mathbf{n}) \cdot \partial F/\partial \mathbf{n} = 0$. Therefore the directions $\partial F/\partial \mathbf{n}$ form the envelope of planes perpendicular to $\mathbf{n}$, as required. The directions $\partial F/\partial \mathbf{n}$ at a point are called bi-characteristic directions, as illustrated in Figure 5.3.1a.

A curve that lies on a characteristic surface, and for which the tangent direction is everywhere a bi-characteristic direction on the local Monge cone, is

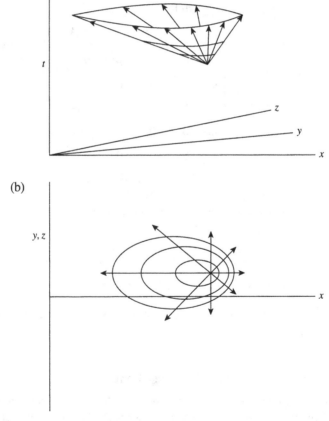

Fig. 5.3.1. (a) Monge cone, either infinitesimal or a tangent surface, and its bi-characteristic directions. (b) Projection of (a) to (x, y, z) space, showing wavefronts and rays in an infinitesimal region.

called a bi-characteristic curve, or simply a bi-characteristic. An arbitrary point (t, x, y, z) on a bi-characteristic may be projected to (x, y, z) for successive values of t, to determine the motion of a point on a curve in three-dimensional (x, y, z) space. This curve is called a ray, and sections of the characteristic surface at successive values of t are surfaces in (x, y, z) space called wavefronts. Therefore to characteristic surfaces and bi-characteristic curves in the four-dimensional space (t, x, y, z) there correspond wavefronts and rays in the three-dimensional space (x, y, z). In general, rays are not perpendicular to wavefronts. The projection to (x, y, z) space is illustrated in Figure 5.3.1b for the special case of the infinitesimal region corresponding to a Monge cone, which could represent the vertex of a ray conoid. Sections of a ray conoid at successive t represent wavefront propagation outwards from a point disturbance.

5.4 Ray Equations

A curve on a characteristic surface $\phi(\mathbf{x}) = 0$, where $\mathbf{x} = (t, x, y, z)$, may be represented parametrically in terms of a parameter s as function $\mathbf{x}(s)$ satisfying $\phi(\mathbf{x}(s)) = 0$ for all s. The normal $\mathbf{n} = (n_0, n_1, n_2, n_3)$ at $\mathbf{x}(s)$ may then be represented as $\mathbf{n}(s)$, as shown in Figure 5.4.1. We now derive the ordinary differential

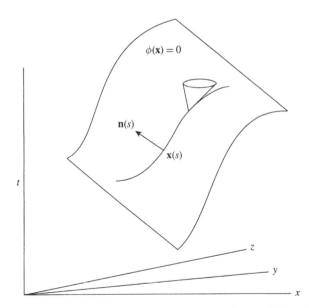

Fig. 5.4.1. Bi-characteristic curve $\mathbf{x}(s)$ and normal $\mathbf{n}(s)$ on a characteristic surface $\phi(\mathbf{x}) = 0$. Also shown is a Monge cone on the bi-characteristic curve. The vectors $\mathbf{x}$ and $\mathbf{n}$ are four-dimensional, that is, they have components (t, x, y, z) and (n_0, n_1, n_2, n_3).

equations satisfied by $\mathbf{x}(s)$ and $\mathbf{n}(s)$ when the curve is a bi-characteristic, that is, when it is tangential at each point to the local Monge cone.

Since the direction rather than the magnitude of $\mathbf{n}$ is of interest, we may take

$$n_k = \frac{\partial \phi}{\partial x_k}. \tag{5.4.1}$$

Then

$$\frac{\partial n_k}{\partial x_l} = \frac{\partial^2 \phi}{\partial x_k \partial x_l} = \frac{\partial n_l}{\partial x_k}. \tag{5.4.2}$$

We shall include in the notation the dependence of the cone of normals on position $\mathbf{x}$, and so we write the equation of the cone of normals as

$$F(\mathbf{x}, \mathbf{n}) = 0. \tag{5.4.3}$$

An infinitesimal segment $d\mathbf{x}$ of a bi-characteristic is parallel to $\partial F / \partial \mathbf{n}$. Therefore by redefining s if necessary, we may take

$$\frac{d\mathbf{x}}{ds} = \frac{\partial F}{\partial \mathbf{n}}. \tag{5.4.4}$$

Since the direction $\mathbf{n}$ is a function of position $\mathbf{x}$ on the characteristic surface, differentiation of (5.4.3) gives

$$\frac{\partial F}{\partial x_k} + \frac{\partial F}{\partial n_l} \frac{\partial n_l}{\partial x_k} = 0. \tag{5.4.5}$$

By (5.4.2) and (5.4.4), this is

$$\frac{\partial F}{\partial x_k} + \frac{dx_l}{ds} \frac{\partial n_k}{\partial x_l} = 0. \tag{5.4.6}$$

The chain rule gives

$$\frac{\partial F}{\partial \mathbf{x}} + \frac{d\mathbf{n}}{ds} = 0. \tag{5.4.7}$$

Equations (5.4.4) and (5.4.7) together are

$$\frac{d\mathbf{x}}{ds} = \frac{\partial F}{\partial \mathbf{n}}, \qquad \frac{d\mathbf{n}}{ds} = -\frac{\partial F}{\partial \mathbf{x}}. \tag{5.4.8}$$

These are Hamilton's equations. Thus the function $F(\mathbf{x}, \mathbf{n})$ defining the cone of normals is a Hamiltonian. The equations are consistent with the original equation $F(\mathbf{x}, \mathbf{n}) = 0$ because

$$
\begin{aligned}
\frac{d}{ds} F(\mathbf{x}, \mathbf{n}) &= \frac{\partial F}{\partial \mathbf{x}} \cdot \frac{d\mathbf{x}}{ds} + \frac{\partial F}{\partial \mathbf{n}} \cdot \frac{d\mathbf{n}}{ds} \\
&= -\frac{d\mathbf{n}}{ds} \cdot \frac{d\mathbf{x}}{ds} + \frac{d\mathbf{x}}{ds} \cdot \frac{d\mathbf{n}}{ds} \\
&= 0.
\end{aligned}
\tag{5.4.9}
$$

Hamilton's equations (5.4.8) could be called the bi-characteristic equations, but in fact they are usually called the ray-tracing equations or the ray equations as they also determine the rays. Given any $\mathbf{x}_0$, it is possible to determine the ray conoid with vertex at $\mathbf{x}_0$ by solving the ordinary differential equations (5.4.8) for each initial condition $(\mathbf{x}_0, \mathbf{n}_0)$ for which $\mathbf{n}_0$ satisfies $F(\mathbf{x}_0, \mathbf{n}_0) = 0$.

5.5 Application to High Speed Flow

We now apply the theory of characteristics, bi-characteristics, wavefronts, and rays to the equations of fluid dynamics. In (5.2.18) we defined $m = n_0 + un_1 + vn_2 + wn_3$ and found that the equation of the cone of normals factorises into the separate equations $m^3 = 0$ and $m^2 = c^2(n_1^2 + n_2^2 + n_3^2)$. The former is a triply repeated equation $m = 0$, linear in $\mathbf{n}$, and the latter is a single irreducible equation, quadratic in $\mathbf{n}$.

The equation $n_0 + un_1 + vn_2 + wn_3 = 0$ asserts that $\mathbf{n}$ is perpendicular to $(1, u, v, w)$. Therefore the part of the cone of normals corresponding to $m = 0$ is a plane. Each plane perpendicular to an $\mathbf{n}$ in this plane contains the vector $(1, u, v, w)$, so that the corresponding part of the Monge cone is a line parallel to $(1, u, v, w)$. This line gives the only bi-characteristic direction corresponding to $m = 0$.

To obtain these results by the formulae of Sections 5.2–5.4, let an arbitrary characteristic surface corresponding to $m = 0$ have equation $\phi(\mathbf{x}) = 0$. Replacing n_k by $\partial\phi/\partial x_k$ in $n_0 + un_1 + vn_2 + wn_3 = 0$ gives the differential equation satisfied by ϕ. Since $(x_0, x_1, x_2, x_3) = (t, x, y, z)$, this equation is

$$
\frac{\partial \phi}{\partial t} + u\frac{\partial \phi}{\partial x} + v\frac{\partial \phi}{\partial y} + w\frac{\partial \phi}{\partial z} = 0.
\tag{5.5.1}
$$

Thus $D\phi/Dt = 0$, and corresponding to $m = 0$ there are possible surfaces of nonsmoothness that propagate "with the fluid."

The bi-characteristic directions corresponding to $m = 0$ are $\partial m/\partial \mathbf{n} = (\partial/\partial \mathbf{n})(n_0 + un_1 + vn_2 + wn_3) = (1, u, v, w)$, as we found by geometrical reasoning. An infinitesimal segment $d\mathbf{x} = (dt, dx, dy, dz)$ of a bi-characteristic curve satisfies

$$(dt, dx, dy, dz) \propto (1, u, v, w). \tag{5.5.2}$$

Therefore

$$\left(\frac{dx}{dt}, \frac{dy}{dt}, \frac{dz}{dt}\right) = (u, v, w). \tag{5.5.3}$$

This is a ray version of (5.5.1). It expresses the fact that rays corresponding to $m = 0$ describe propagation at the velocity of the fluid. The quantities that so propagate can be determined by analysis of the partial differential equations. We note only the conclusion, which is that corresponding to $m = 0$ is a possible vortex sheet or entropy jump. This is consistent with the cubed factor m in the equation of the cone of normals, because two factors correspond to the tangential components of velocity, which can jump at a vortex sheet, and the remaining factor corresponds to the entropy jump.

The quadratic factor in the equation of the cone of normals is $(n_0 + un_1 + vn_2 + wn_3)^2 = c^2(n_1^2 + n_2^2 + n_3^2)$, so that the corresponding equation for a characteristic surface $\phi(\mathbf{x}) = 0$ is

$$\left(\frac{\partial \phi}{\partial t} + u\frac{\partial \phi}{\partial x} + v\frac{\partial \phi}{\partial y} + w\frac{\partial \phi}{\partial z}\right)^2 = c^2\left\{\left(\frac{\partial \phi}{\partial x}\right)^2 + \left(\frac{\partial \phi}{\partial y}\right)^2 + \left(\frac{\partial \phi}{\partial z}\right)^2\right\}.$$
$$\tag{5.5.4}$$

That is,

$$\left|\frac{D\phi}{Dt}\right| = c|\nabla\phi|. \tag{5.5.5}$$

Therefore the wavefronts propagate relative to the fluid at the local speed of sound. To show how the ray directions may be found from the theory of Sections 5.2–5.4, we put $m = m(\mathbf{n}) = n_0 + un_1 + vn_2 + wn_3$ and define $F(\mathbf{n})$ by

$$F(\mathbf{n}) = m^2 - c^2(n_1^2 + n_2^2 + n_3^2). \tag{5.5.6}$$

The corresponding part of the Monge cone is generated by the bi-characteristic directions

$$\frac{\partial F}{\partial \mathbf{n}} = 2(m, mu - c^2n_1, mv - c^2n_2, mw - c^2n_3). \tag{5.5.7}$$

Therefore infinitesimal segments of bi-characteristic curves satisfy

$$(dt, dx, dy, dz) \propto \left(m, mu - c^2 n_1, mv - c^2 n_2, mw - c^2 n_3\right). \qquad (5.5.8)$$

Hence

$$\left(\frac{dx}{dt}, \frac{dy}{dt}, \frac{dz}{dt}\right) = \left(u - \frac{c^2 n_1}{m}, v - \frac{c^2 n_2}{m}, w - \frac{c^2 n_3}{m}\right). \qquad (5.5.9)$$

Thus

$$\left(\frac{dx}{dt} - u\right)^2 + \left(\frac{dy}{dt} - v\right)^2 + \left(\frac{dz}{dt} - w\right)^2 = \frac{c^4\left(n_1^2 + n_2^2 + n_3^2\right)}{m^2}. \qquad (5.5.10)$$

Since $m^2 = c^2(n_1^2 + n_2^2 + n_3^2)$, we obtain

$$\left(\frac{dx}{dt} - u\right)^2 + \left(\frac{dy}{dt} - v\right)^2 + \left(\frac{dz}{dt} - w\right)^2 = c^2. \qquad (5.5.11)$$

This is a ray version of (5.5.4). The propagating quantity could be a small acoustic disturbance or a jump in a derivative of the pressure in the direction normal to a wavefront.

At a point (x, y, z), the ray directions satisfying (5.5.11) may be found by drawing an arrow to represent (u, v, w), and then drawing a sphere of radius c around the tip of the arrow, as shown in Figure 5.5.1. The ray directions are from (x, y, z) to the surface of the sphere. For subsonic flow (i.e., $u^2 + v^2 + w^2 < c^2$), the point (x, y, z) is inside the sphere, and the ray directions surround the point completely. For supersonic flow (i.e., $u^2 + v^2 + w^2 > c^2$), the point (x, y, z) is outside the sphere, and the ray directions fill a cone. The two cases reflect the fact that in the space (n_0, n_1, n_2, n_3) the equation $(n_0 + un_1 + vn_2 + wn_3)^2 = c^2(n_1^2 + n_2^2 + n_3^2)$ gives an elliptic cone when $u^2 + v^2 + w^2 < c^2$, and a hyperbolic cone when $u^2 + v^2 + w^2 > c^2$. This is readily illustrated for flow in two space dimensions, for which the cone of normals is

$$(n_0 + un_1 + vn_2)^2 = c^2\left(n_1^2 + n_2^2\right). \qquad (5.5.12)$$

Let us put $n_0 = 1$, which is permissible because the equation is homogeneous in n_0, n_1, n_2. Then

$$(1 + un_1 + vn_2)^2 = c^2\left(n_1^2 + n_2^2\right). \qquad (5.5.13)$$

When $u^2 + v^2 < c^2$, the values (n_1, n_2) lie on an ellipse, and the corresponding

(a)

(b)

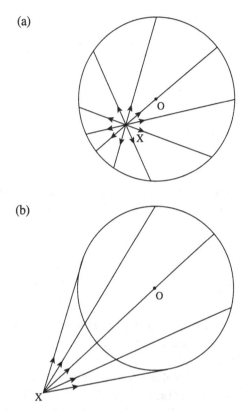

Fig. 5.5.1. Ray directions from a point $X = (x, y, z)$. The vector XO represents the velocity (u, v, w), and the radius of the sphere centred on O is the speed of sound c at X. (a) Subsonic flow, $u^2 + v^2 + w^2 < c^2$. (b) Supersonic flow, $u^2 + v^2 + w^2 > c^2$.

(n_0, n_1, n_2) lie on an elliptic cone, as in Figure 5.5.2a. Similarly, when $u^2 + v^2 > c^2$, the values (n_1, n_2) lie on a hyperbola, and the corresponding (n_0, n_1, n_2) lie on a hyperbolic cone, as in Figure 5.5.2b. By considering the envelope of planes perpendicular to $\mathbf{n}$, we may visualise the corresponding Monge cones and rays.

Let the jump of a nonsmooth quantity across a characteristic surface be Δ. For example, Δ might be the jump in a derivative of the pressure. The values of Δ at different points on the same ray are related by an ordinary differential equation, called the transport equation for Δ, obtainable from the original partial differential equations. Thus if position on a ray is parameterised by s, the transport equation gives an expression for $d\Delta/ds$ in terms of Δ and other variables. An extensive mathematical theory of transport equations has been developed, much of which is relevant to high speed flow, but it is beyond the scope of this book.

(a)

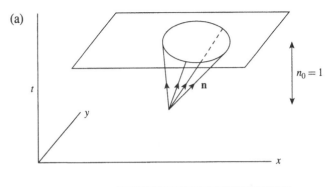

(b)

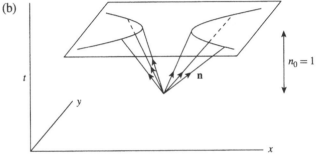

Fig. 5.5.2. Cones of normals for two-dimensional flow. (a) Elliptic cone, subsonic flow. (b) Hyperbolic cone, supersonic flow.

5.6 Bibliographic Notes

Accounts of the mathematical theory of characteristic surfaces and rays, with particular reference to the equations of fluid dynamics, are given in many of the works in Tables 13.3.1 and 13.4.1, for example Courant & Friedrichs (1948), Ferri (1955a), Friedrichs (1955), and Meyer (1953, 1960). See also Friedlander (1958), Courant & Hilbert (1962), Prasad & Ravindram (1985), and Prasad (1993). The most important early paper is Riemann (1860). A paper on transport equations is Varley & Cumberbatch (1965).

5.7 Further Results and Exercises

1. The equations of motion for two-dimensional flow of an ideal fluid may be written in cylindrical coordinates (r, θ) as

$$\rho_t + \frac{1}{r}\{(r\rho u)_r + (\rho v)_\theta\} = 0,$$

$$u_t + uu_r + \frac{vu_\theta}{r} - \frac{v^2}{r} + \frac{p_r}{\rho} = 0,$$

$$v_t + uv_r + \frac{vv_\theta}{r} + \frac{uv}{r} + \frac{p_\theta}{\rho r} = 0,$$

$$s_t + us_r + \frac{v}{r}s_\theta = 0.$$

The vector $\mathbf{n} = (n_0, n_1, n_2)$ is perpendicular to a surface in (t, r, θ) space. Determine, in terms of $\mathbf{n}$, the condition for the surface to be a characteristic surface of the equations of motion. Hence obtain the equations for the two families of ray directions. Let the speed of sound be c. Show that the rays of one family all satisfy the relation

$$\left(\frac{dr}{dt} - u \right)^2 + \left(r\frac{d\theta}{dt} - v \right)^2 = c^2.$$

Determine the corresponding relation for the other family. Comment briefly on the physical significance of the two types of rays.

2. Explain what is meant by a characteristic surface of a system of partial differential equations. Illustrate your remarks with reference to the equations of isentropic fluid-dynamical flow in one dimension, for which the equations of motion are

$$\frac{\partial u}{\partial t} + u\frac{\partial u}{\partial x} + \frac{c^2}{\rho}\frac{\partial \rho}{\partial x} = 0,$$

$$\frac{\partial \rho}{\partial t} + u\frac{\partial \rho}{\partial x} + \rho\frac{\partial u}{\partial x} = 0.$$

3. Discuss the theory of characteristic surfaces for the steady homentropic flow of an ideal fluid, deriving any formulae you use from the steady flow equations. How could your results be obtained from the corresponding results for unsteady flow?

6

Shocks

6.1 Jump Conditions

Throughout this chapter we consider a fluid that outside of a thin layer is ideal, that is, may be assumed to have no viscosity or thermal conductivity and to have stress tensor σ_{ij} given in terms of the pressure p by $\sigma_{ij} = -p\delta_{ij}$. We represent the thin layer as a surface of discontinuity, and we apply the jump conditions (2.1.11), (2.2.8), and (2.3.5) with $\mathbf{q} = 0$ and $\mathbf{V} = 0$ (i.e., in a frame of reference moving with the surface). We also use the condition that entropy cannot decrease. Then the jump conditions for mass, momentum, energy, and entropy are

$$[\rho\mathbf{u} \cdot \mathbf{n}] = 0, \tag{6.1.1a}$$

$$[p\mathbf{n} + \rho\mathbf{u}(\mathbf{u} \cdot \mathbf{n})] = 0, \tag{6.1.1b}$$

$$\left[\left(p + \rho e + \tfrac{1}{2}\rho\mathbf{u}^2\right)(\mathbf{u} \cdot \mathbf{n})\right] = 0, \tag{6.1.1c}$$

$$[s] \geq 0. \tag{6.1.1d}$$

We use coordinates (x, y, z) in which the surface of discontinuity is perpendicular to the x axis, and the fluid velocity $\mathbf{u} = (u, v, w)$ satisfies $u \geq 0$, as shown in Figure 6.1.1. Thus before reaching the surface, the fluid is on the upstream side $x < 0$, which we shall call the front of the surface, or side 1, and afterwards the fluid is on the downstream side $x > 0$, that is, the back of the surface, or side 2. Accordingly, subscripts 1 and 2 will be used to indicate the side on which a quantity is evaluated. In components, the jump conditions (6.1.1) are

$$[\rho u] = 0,$$
$$[p + \rho u^2] = 0, \qquad [\rho uv] = 0, \qquad [\rho uw] = 0,$$
$$\left[\left(p + \rho e + \tfrac{1}{2}\rho\mathbf{u}^2\right)u\right] = 0,$$
$$[s] \geq 0. \tag{6.1.2}$$

85

6 Shocks

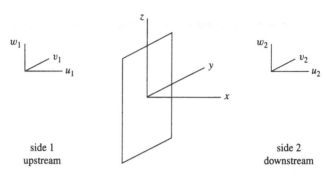

side 1   side 2
upstream   downstream

Fig. 6.1.1. Coordinate system (x, y, z) in which a surface of discontinuity is at rest and
perpendicular to the x axis. Upstream velocity (u_1, v_1, w_1) and downstream velocity
(u_2, v_2, w_2) (i.e., $u_1 \geq 0$ and $u_2 \geq 0$).

First assume that $u_1 = 0$. The condition $[\rho u] = 0$ (i.e., $\rho_1 u_1 = \rho_2 u_2$) and
the fact that ρ_1 and ρ_2 are nonzero give $u_2 = 0$. Therefore $[p + \rho u^2] = 0$ becomes $[p] = 0$, that is, $p_1 = p_2$. The remaining jump conditions are satisfied
by arbitrary values of v_1, v_2 and w_1, w_2. Since no fluid is passing through the
surface, the requirement of nondecreasing entropy becomes irrelevant, and we
may allow arbitrary values of s_1 and s_2. Thus $u_1 = 0$ gives a surface of discontinuity with zero normal components of velocity, continuous pressure, arbitrary
tangential components of velocity, and arbitrary entropy. Such a surface is a
vortex sheet, also called a contact discontinuity, contact surface, slip surface,
or slip stream. Photographs and diagrams usually show a section of (x, y, z)
space, and the surface appears as a line, called a contact line or slip line.

Now assume that $u_1 \neq 0$. The condition $\rho_1 u_1 = \rho_2 u_2$, with nonzero ρ_1 and
ρ_2, gives $u_2 \neq 0$. Therefore the conditions $[\rho u v] = 0$ and $[\rho u w] = 0$ (i.e.,
$\rho_1 u_1 v_1 = \rho_2 u_2 v_2$ and $\rho_1 u_1 w_1 = \rho_2 u_2 w_2$) may be divided by $\rho_1 u_1 = \rho_2 u_2$
to give $v_1 = v_2$ and $w_1 = w_2$. Thus the tangential components of velocity
are continuous. By continuity of ρu, v^2, and w^2 it follows that $[(\rho \mathbf{u}^2)u] =
[(\rho u^2)u]$. Therefore the condition $[(p + \rho e + \frac{1}{2}\rho \mathbf{u}^2)u] = 0$, written in terms
of specific enthalpy $h = e + p/\rho$, is $[(h + \frac{1}{2}u^2)\rho u] = 0$, and on dividing by
the continuous quantity ρu we obtain $[h + \frac{1}{2}u^2] = 0$. Thus $u_1 \neq 0$ gives a
surface of discontinuity with continuous tangential components of velocity and
continuous values of ρu, $p + \rho u^2$, and $h + \frac{1}{2}u^2$. Such a surface is a shock wave
(i.e., a shock). Thus the shock conditions, in their simplest form, are

$$[\rho u] = 0, \qquad [p + \rho u^2] = 0, \qquad \left[h + \tfrac{1}{2}u^2\right] = 0,$$
$$[s] \geq 0, \qquad [v] = 0, \qquad [w] = 0. \tag{6.1.3}$$

Since there are no other possibilities besides $u_1 = 0$ and $u_1 \neq 0$, the above results show that it is impossible to "mix" a vortex sheet and a shock. That is, at a vortex sheet there cannot be a jump in normal velocity, and at a shock there cannot be a jump in tangential velocity.

6.2 Normal Shocks in an Arbitrary Fluid.
The Rankine–Hugoniot Relation

We now derive some algebraic consequences of the jump conditions for a shock that is perpendicular to the flow in front and behind, that is, for a normal shock. Thus the tangential components of velocity are zero, and the problem is one-dimensional. We take as given the conditions in front of the shock (on side 1), and we denote the average of conditions on sides 1 and 2 by a bar, so that, for example, the average specific volume is $\bar{v} = \frac{1}{2}(v_1 + v_2) = \frac{1}{2}(1/\rho_1 + 1/\rho_2)$. We take the independent thermodynamic variables to be p and v, and since we are considering an arbitrary fluid we express all results in terms of general functions $h(p, v)$, $s(p, v)$, and specific internal energy $e(p, v)$.

The jump conditions determine a relation between the pressures p_1 and p_2, specific volumes v_1 and v_2, and enthalpies $h_1 = h(p_1, v_1)$ and $h_2 = h(p_2, v_2)$. To obtain this relation, we start with $[\rho u] = 0$ and define the mass flux m by

$$m = \rho_1 u_1 = \rho_2 u_2. \tag{6.2.1}$$

Then $[p + \rho u^2] = 0$ becomes $[p + m^2/\rho] = 0$, that is,

$$[p] + m^2[v] = 0. \tag{6.2.2}$$

Similarly, $[h + \frac{1}{2}u^2] = 0$ becomes $[h + \frac{1}{2}m^2/\rho^2] = 0$, that is,

$$[h] + \frac{1}{2}m^2[v^2] = 0. \tag{6.2.3}$$

Equation (6.2.2) is

$$m^2 = -\frac{[p]}{[v]} = -\frac{p_2 - p_1}{v_2 - v_1}. \tag{6.2.4}$$

Substituting (6.2.4) into (6.2.3), and using $[v^2]/[v] = (v_2^2 - v_1^2)/(v_2 - v_1) = v_2 + v_1 = 2\bar{v}$, we obtain

$$[h] = \frac{1}{2}\frac{[v^2][p]}{[v]} = \bar{v}[p]. \tag{6.2.5}$$

In full,

$$h(p_2, v_2) - h(p_1, v_1) = \tfrac{1}{2}(v_1 + v_2)(p_2 - p_1). \tag{6.2.6}$$

This is the Rankine–Hugoniot relation. If p_1 and v_1 are fixed, it gives the Rankine–Hugoniot adiabatic, or shock adiabatic, relating p_2 and v_2, that is, relating the pressure and specific volume of the fluid after passing through the shock. Hence for each (p_1, v_1) we may draw a Rankine–Hugoniot curve in the (p, v) plane specifying the possible downstream thermodynamic states. The position of (p_2, v_2) on this curve is determined by the strength of the shock. Since (6.2.6) is always satisfied by $(p_2, v_2) = (p_1, v_1)$, the curve always passes through the point $(p, v) = (p_1, v_1)$, corresponding to the limiting case of a shock of zero strength.

A typical Rankine–Hugoniot curve is shown in Figure 6.2.1. The slope of the chord between (p_1, v_1) and (p_2, v_2) is $(p_2 - p_1)/(v_2 - v_1)$, which by (6.2.4) is $-m^2$. The value of m determines the strength of the shock. Thus if p_1, v_1, and m are given, the construction of a chord of slope $-m^2$ through (p_1, v_1) on the Rankine–Hugoniot curve determines (p_2, v_2). Hence the curve gives an easily visualised relation between conditions before and after fluid has passed through a normal shock.

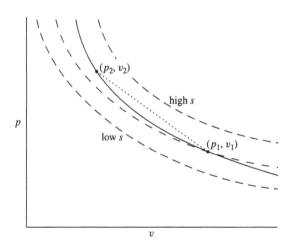

Fig. 6.2.1. The Rankine–Hugoniot curve (——) for a normal shock, and its relation to constant-entropy curves (– – –) for a fluid with a convex equation of state (i.e., with a positive coefficient of nonlinearity Γ) and with a positive coefficient of thermal expansion (i.e., with a positive Grüneisen parameter G). The fluid before and after passing through the shock is represented by points (p_1, v_1) and (p_2, v_2), and the chord (·····) joining these points has slope $-m^2 = [p]/[v] = -(\rho_1 u_1)^2 = -(\rho_2 u_2)^2$.

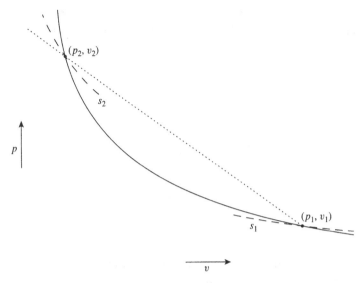

Fig. 6.2.2. Expanded view of Figure 6.2.1, showing that at (p_1, v_1) the Rankine–Hugoniot curve and the constant-entropy curve have a common tangent and that at (p_2, v_2) the constant-entropy curve lies between the Rankine–Hugoniot curve and the chord to (p_1, v_1).

In most conditions, fluids have a positive coefficient of thermal expansion (i.e, $\beta > 0$) and except close to a critical point have a convex equation of state (i.e., $(\partial^2 p/\partial v^2)_s > 0$). We saw after (3.4.5) and (3.2.1) that the Grüneisen parameter G and the coefficient of nonlinearity Γ are then positive, and we shall assume this henceforth. It can be deduced that constant-entropy curves are then oriented relative to the Rankine–Hugoniot curve and the chord joining (p_1, v_1) and (p_2, v_2) as shown in Figures 6.2.1 and 6.2.2. The entropy $s(p, v)$ increases as (p, v) moves on the Rankine–Hugoniot curve upwards and to the left. Thus $[s] > 0$ implies that (p_2, v_2) is higher than and left of (p_1, v_1) (i.e., that $p_2 > p_1$ and $v_2 < v_1$). From (6.2.6), and the relations $\rho = 1/v$ and $\rho_1 u_1 = \rho_2 u_2$, there follows $h_2 > h_1$, $\rho_2 > \rho_1$, and $u_2 < u_1$. It can also be deduced that the internal energy increases (i.e., $e_2 > e_1$).

We now concentrate on velocities and determine the Mach number on each side of the shock. The definition $m = \rho_1 u_1 = \rho_2 u_2$ gives $u_1 = m/\rho_1 = mv_1$ and $u_2 = m/\rho_2 = mv_2$, so that

$$[u] = m[v]. \tag{6.2.7}$$

Note that m, u_1, and u_2 are positive, by definition of side 1 as the upstream side of the shock, and also that we have just shown $[u]$ and $[v]$ to be negative. Eliminating m from $m^2 = -[p]/[v]$ and $[u] = m[v]$, giving due regard to signs, we obtain

$$-[u] = m(-[v]) = \left(\frac{[p]}{-[v]} \right)^{\frac{1}{2}} (-[v]) = (-[p][v])^{\frac{1}{2}}. \qquad (6.2.8)$$

On each side of the shock, the speed of sound c satisfies

$$c^2 = \left(\frac{\partial p}{\partial \rho} \right)_s = -v^2 \left(\frac{\partial p}{\partial v} \right)_s = -\frac{u^2}{m^2} \left(\frac{\partial p}{\partial v} \right)_s, \qquad (6.2.9)$$

so that the Mach number $M = u/c$ satisfies

$$\frac{1}{M^2} = -\frac{1}{m^2} \left(\frac{\partial p}{\partial v} \right)_s = \left(\frac{\partial p}{\partial v} \right)_s \Big/ \left(\frac{[p]}{[v]} \right). \qquad (6.2.10)$$

Evaluation of (6.2.10) at (p_1, v_1) and (p_2, v_2) determines the Mach numbers $M_1 = u_1/c_1$ and $M_2 = u_2/c_2$ in front of and behind the shock. Since in Figure 6.2.2 the dashed lines have slope $(\partial p / \partial v)_s$ and the dotted line has slope $[p]/[v]$, it follows from the orientation of the lines that $1/M_1^2 < 1$ and $1/M_2^2 > 1$, that is,

$$M_1 > 1, \qquad M_2 < 1. \qquad (6.2.11)$$

An observer moving with the fluid on side 1 sees the shock approaching, and an observer moving with the fluid on side 2 sees the shock receding. Therefore we have shown that a shock approaches supersonically and recedes subsonically.

The Rankine–Hugoniot relation takes a simple limiting form for a weak shock. The points (p_1, v_1) and (p_2, v_2) are then close together, so that the limiting form is obtained by performing a Taylor series expansion of the relation and keeping only the leading terms. We shall omit the suffix 1 from the variables defined on side 1 of the shock, so that (p, v) and (p_2, v_2) are neighbouring points and the Rankine–Hugoniot relation is

$$h_2 - h = \tfrac{1}{2}(v + v_2)(p_2 - p). \qquad (6.2.12)$$

We take as independent variables the pressure and entropy, and we denote their increments as the fluid passes through the shock by δp and δs, so that

$$\delta p = p_2 - p, \qquad \delta s = s_2 - s. \qquad (6.2.13)$$

Our starting point is the thermodynamic relation

$$dh = T ds + v dp. \qquad (6.2.14)$$

With partial derivatives denoted by subscripts p and s, this relation gives

$$h_p = v, \qquad h_{pp} = v_p, \qquad \cdots,$$
$$h_s = T, \qquad h_{ss} = T_s, \qquad \cdots. \qquad (6.2.15)$$

Therefore

$$h_2 = h + h_s \delta s + h_p \delta p + \cdots$$
$$= h + T \delta s + v \delta p + \tfrac{1}{2} v_p (\delta p)^2 + \tfrac{1}{6} v_{pp} (\delta p)^3 + \cdots, \qquad (6.2.16)$$
$$v_2 = v + v_s \delta s + v_p \delta p + \tfrac{1}{2} v_{pp} (\delta p)^2 + \cdots.$$

Substitution of (6.2.16) into (6.2.12), followed by cancellation of the terms in $v \delta p$ and $v_p (\delta p)^2$, gives

$$T \delta s + \tfrac{1}{6} v_{pp} (\delta p)^3 + \cdots = \left(\tfrac{1}{2} v_s \delta s + \tfrac{1}{4} v_{pp} (\delta p)^2 + \cdots \right) \delta p. \qquad (6.2.17)$$

In this equation, $\tfrac{1}{2} v_s \delta s \delta p$ is a higher order differential than $T \delta s$ and may be ignored. The remaining terms can balance only if δs and $(\delta p)^3$ are of the same order. The dominant terms in (6.2.17) then give

$$T \delta s \simeq \left(\tfrac{1}{4} - \tfrac{1}{6} \right) v_{pp} (\delta p)^3, \qquad (6.2.18)$$

or

$$\delta s \simeq \frac{1}{12} \frac{1}{T} v_{pp} (\delta p)^3. \qquad (6.2.19)$$

As δs and $(\delta p)^3$ are of the same order, it follows that the terms omitted in (6.2.16) (e.g., terms of order $(\delta s)^2$) are negligible.

Equation (6.2.19) implies that the change in entropy at a weak shock may be ignored. Indeed, the change may often be ignored when the shock is of moderate strength. One consequence is that a shock generates only a small amount of vorticity unless it is highly curved or the upstream flow is nonuniform. Another consequence of (6.2.19) is that at (p_1, v_1) the constant-entropy curve has second-order contact with the Rankine–Hugoniot curve, that is, in Figure 6.2.3 we have BC $\propto$ (AB)^3. Therefore $[p]/[v]$ is close to $(\partial p / \partial v)_s$, so that, by (6.2.10), a weak shock travels at close to the speed of sound: It is slightly supersonic relative to the fluid in front, and it is slightly subsonic relative to the fluid behind.

The extension of the theory in this section to flows in which the coefficient of nonlinearity Γ or the Grüneisen parameter G are negative is a complicated matter, because the Rankine–Hugoniot curve and the curves of constant s (i.e., the isentropes) need not then be as shown in Figures 6.2.1–6.2.3. For example, if $G < 0$ (or equivalently $\beta < 0$) then the curve s_2 through (p_2, v_2) in Figure 6.2.2 no longer slopes as shown but is steeper than the Rankine–Hugoniot curve. In

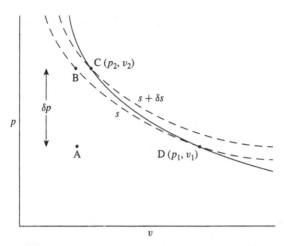

Fig. 6.2.3. Definitions of δs and δp when Figure 6.2.1 is used for a weak shock. Since $\delta s \propto$ BC and $\delta p = $ AB, the weak shock relation $\delta s \propto (\delta p)^3$ gives BC $\propto$ (AB)^3, which implies that the Rankine–Hugoniot curve CD and the constant-entropy curve BD have second-order contact at D.

real gases, the Rankine–Hugoniot curve may have an S-shaped part, with points at which the tangent is vertical, and theoretical conditions are even required to exclude the possibility of horizontal tangents or isolated closed loops. The most important theoretical result is the Bethe–Weyl theorem, which asserts that the Rankine–Hugoniot curve through any point (p_1, v_1) (at which the entropy is s_1) intersects every isentrope at least once, and, if $\Gamma > 0$ everywhere on an isentrope s_2, then the Rankine–Hugoniot curve through (p_1, v_1) intersects this isentrope exactly once (say at (p_2, v_2)); in this case, $v_2 < v_1$ if $s_2 > s_1$, and $v_2 > v_1$ if $s_2 < s_1$. The original proofs of the theorem depended on assumptions about the value of G, but the theorem has since been proved to hold irrespective of G. Further analysis shows that if Γ is negative in some thermodynamic states (i.e., if the fluid does not have a convex equation of state) then it is possible to have $v_2 > v_1$ when $s_2 > s_1$; that is, a rarefaction shock is possible. In the laboratory, rarefaction shocks may be produced in the so-called "dense gases" or "Bethe–Zel'dovich–Thompson fluids." The relation between thermodynamics and shock behavior, for example shock-splitting and stability, is an active area of research.

6.3 Normal Shocks in a Polytropic Gas. The Prandtl Relation

We now apply the results of the previous section to a perfect gas with constant specific heats (i.e., to a polytropic gas). The variables we shall use are pressure

p, density ρ, temperature T, entropy s, enthalpy h, speed of sound c, flow speed u, and Mach number M, and the properties specifying the gas are its gas constant R, the constant specific heats c_p and c_v, and the ratio of specific heats $\gamma = c_p/c_v$. Thus γ is a constant and

$$p = \rho RT, \qquad p \propto e^{s/c_v}\rho^{\gamma}, \qquad c_v = \frac{R}{\gamma - 1},$$

$$h = \frac{\gamma RT}{\gamma - 1}, \qquad c^2 = \gamma RT, \qquad u = Mc. \tag{6.3.1}$$

Our algebraic technique will be to work with ρ, M, and c and obtain the relation between the Mach numbers M_1 and M_2 on sides 1 and 2 of the shock. The algebra in the theory of shocks can never be entirely trivial, because the jump conditions are homogeneous and so admit solutions in which the jumps are all zero, that is, in which there is no shock because conditions on sides 1 and 2 are the same. Thus a general algebraic relation always contains a factor that equals zero when the shock is absent and is consequently of no interest, while the information about the shock resides in the other factor. For example, the relation between M_1 and M_2 contains a factor $M_2^2 - M_1^2$ that must be removed, and a comparable step is required no matter which algebraic method is adopted. Thus information about shocks is "coded" in expressions of the form $(f(a) - f(b))/(b - a)$. A feature of solving problems involving jump conditions is that some skill is needed to limit the complexity of the intermediate stages of the algebra.

To express the jump conditions $[\rho u] = 0$, $[p + \rho u^2] = 0$, and $[h + \frac{1}{2}u^2] = 0$ in terms of ρ, M, and c, we use $c^2 = \gamma p/\rho$ and $h = c^2/(\gamma - 1)$ to write $\rho u = \rho M c$, $p + \rho u^2 = \rho c^2/\gamma + \rho M^2 c^2$, and $h + \frac{1}{2}u^2 = c^2/(\gamma - 1) + \frac{1}{2}M^2 c^2$. Thus

$$[M\rho c] = 0, \tag{6.3.2a}$$

$$[(1 + \gamma M^2)\rho c^2] = 0, \tag{6.3.2b}$$

$$\left[\left(1 + \tfrac{1}{2}(\gamma - 1)M^2\right)c^2\right] = 0. \tag{6.3.2c}$$

To eliminate ρ and c (i.e., ρ_1, ρ_2, c_1, and c_2) we use equations (a), (b), and (c) in (6.3.2) to form the combination $(b)^2/\{(a)^2(c)\}$. Thus

$$\left[\frac{(1 + \gamma M^2)^2}{M^2\left(1 + \frac{1}{2}(\gamma - 1)M^2\right)}\right] = 0, \tag{6.3.3}$$

or

$$\frac{\left(1 + \gamma M_1^2\right)^2}{M_1^2\left(1 + \frac{1}{2}(\gamma - 1)M_1^2\right)} = \frac{\left(1 + \gamma M_2^2\right)^2}{M_2^2\left(1 + \frac{1}{2}(\gamma - 1)M_2^2\right)}. \tag{6.3.4}$$

Cross-multiplying and removing a factor $M_2^2 - M_1^2$ gives

$$1 + \frac{1}{2}(\gamma - 1)\left(M_1^2 + M_2^2\right) - \gamma M_1^2 M_2^2 = 0. \tag{6.3.5}$$

Therefore

$$M_2^2 = \frac{1 + \frac{1}{2}(\gamma - 1)M_1^2}{\gamma M_1^2 - \frac{1}{2}(\gamma - 1)}. \tag{6.3.6}$$

The relation between ρ_1 and ρ_2 is given by the combination (b)/(c) in (6.3.2):

$$\frac{\left(1 + \gamma M_1^2\right)\rho_1}{1 + \frac{1}{2}(\gamma - 1)M_1^2} = \frac{\left(1 + \gamma M_2^2\right)\rho_2}{1 + \frac{1}{2}(\gamma - 1)M_2^2}. \tag{6.3.7}$$

Putting $p = \rho c^2/\gamma$ in (6.3.2b) gives

$$\left(1 + \gamma M_1^2\right)p_1 = \left(1 + \gamma M_2^2\right)p_2. \tag{6.3.8}$$

Equation (6.3.2c) is

$$\left(1 + \frac{1}{2}(\gamma - 1)M_1^2\right)^{\frac{1}{2}}c_1 = \left(1 + \frac{1}{2}(\gamma - 1)M_2^2\right)^{\frac{1}{2}}c_2. \tag{6.3.9}$$

Therefore $u = Mc$ gives

$$\frac{\left(1 + \frac{1}{2}(\gamma - 1)M_1^2\right)^{\frac{1}{2}}u_1}{M_1} = \frac{\left(1 + \frac{1}{2}(\gamma - 1)M_2^2\right)^{\frac{1}{2}}u_2}{M_2}, \tag{6.3.10}$$

and $c^2 = \gamma R T$ gives

$$\left(1 + \frac{1}{2}(\gamma - 1)M_1^2\right)T_1 = \left(1 + \frac{1}{2}(\gamma - 1)M_2^2\right)T_2. \tag{6.3.11}$$

From (6.3.1), the entropy jump is

$$s_2 - s_1 = \frac{R}{\gamma - 1}\ln\left\{\left(\frac{p_2}{p_1}\right)\left(\frac{\rho_2}{\rho_1}\right)^{-\gamma}\right\}. \tag{6.3.12}$$

With $v = 1/\rho$ and $h = (\gamma/(\gamma - 1))p/\rho$, the Rankine–Hugoniot relation $h_2 - h_1 = \frac{1}{2}(v_1 + v_2)(p_2 - p_1)$ is

$$\frac{\gamma}{\gamma - 1}\left(\frac{p_2}{\rho_2} - \frac{p_1}{\rho_1}\right) = \frac{1}{2}\left(\frac{1}{\rho_1} - \frac{1}{\rho_2}\right)(p_2 - p_1), \tag{6.3.13}$$

or

$$\frac{\rho_2}{\rho_1} = \frac{(\gamma - 1)p_1 + (\gamma + 1)p_2}{(\gamma + 1)p_1 + (\gamma - 1)p_2}. \tag{6.3.14}$$

Equations (6.3.7)–(6.3.12) may be expressed in terms of M_1 using (6.3.6) and its consequences

$$1 + \tfrac{1}{2}(\gamma - 1)M_2^2 = \frac{\tfrac{1}{4}(\gamma + 1)^2 M_1^2}{\gamma M_1^2 - \tfrac{1}{2}(\gamma - 1)},$$

$$1 + \gamma M_2^2 = \frac{\tfrac{1}{2}(\gamma + 1)\left(1 + \gamma M_1^2\right)}{\gamma M_1^2 - \tfrac{1}{2}(\gamma - 1)}. \tag{6.3.15}$$

The result is

$$\frac{u_2}{u_1} = \frac{1 + \tfrac{1}{2}(\gamma - 1)M_1^2}{\tfrac{1}{2}(\gamma + 1)M_1^2}, \tag{6.3.16a}$$

$$\frac{p_2}{p_1} = 1 + \frac{2\gamma}{\gamma + 1}\left(M_1^2 - 1\right) = \frac{\gamma M_1^2 - \tfrac{1}{2}(\gamma - 1)}{\tfrac{1}{2}(\gamma + 1)}, \tag{6.3.16b}$$

$$\frac{\rho_2}{\rho_1} = \frac{\tfrac{1}{2}(\gamma + 1)M_1^2}{1 + \tfrac{1}{2}(\gamma - 1)M_1^2}, \tag{6.3.16c}$$

$$\frac{T_2}{T_1} = \frac{\left(1 + \tfrac{1}{2}(\gamma - 1)M_1^2\right)\left(\gamma M_1^2 - \tfrac{1}{2}(\gamma - 1)\right)}{\tfrac{1}{4}(\gamma + 1)^2 M_1^2}, \tag{6.3.16d}$$

$$\frac{c_2}{c_1} = \frac{\left(1 + \tfrac{1}{2}(\gamma - 1)M_1^2\right)^{\frac{1}{2}}\left(\gamma M_1^2 - \tfrac{1}{2}(\gamma - 1)\right)^{\frac{1}{2}}}{\tfrac{1}{2}(\gamma + 1)M_1}, \tag{6.3.16e}$$

$$s_2 - s_1 = \frac{R}{\gamma - 1}\ln\left\{\frac{\gamma M_1^2 - \tfrac{1}{2}(\gamma - 1)}{\tfrac{1}{2}(\gamma + 1)}\left(\frac{1 + \tfrac{1}{2}(\gamma - 1)M_1^2}{\tfrac{1}{2}(\gamma + 1)M_1^2}\right)^{\gamma}\right\}. \tag{6.3.16f}$$

Recall that, by (6.2.11), we always have $M_1 > 1$ and $M_2 < 1$ (i.e., supersonic flow upstream and subsonic flow downstream). A check of the above formulae is that the shock vanishes when $M_1 = 1$, that is, M_2 and the right-hand sides of (6.3.16a–e) are then unity, and the right-hand side of (6.3.16f) is zero.

We now derive an unexpectedly simple equation, known as Prandtl's relation, in u_1 and u_2. The jump condition $[h + \tfrac{1}{2}u^2] = 0$ asserts that $c^2/(\gamma - 1) + \tfrac{1}{2}u^2$ is constant across a shock, and we may express the constant in terms of the speed at which the fluid would be sonic, that is, the critical speed $\hat{c}$. Thus $\hat{c}$ is defined by

$$\frac{c^2}{\gamma - 1} + \frac{u^2}{2} = \frac{\hat{c}^2}{\gamma - 1} + \frac{\hat{c}^2}{2} = \frac{1}{2}\left(\frac{\gamma + 1}{\gamma - 1}\right)\hat{c}^2. \qquad (6.3.17)$$

To obtain an equation containing only speeds, we divide $[p + \rho u^2] = 0$ by $[\rho u] = 0$ to obtain $[p/(\rho u) + u] = 0$, and we use the equation of state $p/\rho = RT = c^2/\gamma$. This gives

$$\left[\frac{1}{\gamma}\frac{c^2}{u} + u\right] = 0. \qquad (6.3.18)$$

Thus we have the independent equations

$$\frac{c_1^2}{\gamma - 1} + \frac{u_1^2}{2} = \frac{c_2^2}{\gamma - 1} + \frac{u_2^2}{2} = \frac{1}{2}\left(\frac{\gamma + 1}{\gamma - 1}\right)\hat{c}^2, \qquad (6.3.19a)$$

$$\frac{1}{\gamma}\frac{c_1^2}{u_1} + u_1 = \frac{1}{\gamma}\frac{c_2^2}{u_2} + u_2. \qquad (6.3.19b)$$

Equation (6.3.19a) gives c_1^2 in terms of u_1^2 and $\hat{c}^2$ and gives c_2^2 in terms of u_2^2 and $\hat{c}^2$. Hence (6.3.19b) gives

$$\frac{1}{\gamma u_1}\left\{\frac{1}{2}(\gamma + 1)\hat{c}^2 - \frac{1}{2}(\gamma - 1)u_1^2\right\} + u_1$$

$$= \frac{1}{\gamma u_2}\left\{\frac{1}{2}(\gamma + 1)\hat{c}^2 - \frac{1}{2}(\gamma - 1)u_2^2\right\} + u_2. \qquad (6.3.20)$$

Collecting terms and dividing the resulting cubic expression by $u_1 - u_2$ gives the promised simple equation

$$u_1 u_2 = \hat{c}^2. \qquad (6.3.21)$$

This is Prandtl's relation. If we define Mach numbers relative to the critical speed by $\hat{M}_1 = u_1/\hat{c}$ and $\hat{M}_2 = u_2/\hat{c}$, it may be written

$$\hat{M}_1 \hat{M}_2 = 1. \qquad (6.3.22)$$

Prandtl's relation is useful in calculations and also in providing geometrical interpretations of the shock conditions.

A useful parameter in many problems is the shock strength z, defined as the ratio of the pressure jump across the shock to the upstream pressure. Thus

$$z = \frac{p_2 - p_1}{p_1} = \frac{2\gamma}{\gamma + 1}\left(M_1^2 - 1\right), \qquad (6.3.23)$$

or

$$M_1^2 = 1 + \left(\frac{\gamma + 1}{2\gamma}\right)z. \tag{6.3.24}$$

Therefore (6.3.6) and (6.3.15) become

$$M_2^2 = \frac{1 + \left(\frac{\gamma-1}{2\gamma}\right)z}{1 + z},$$

$$1 + \frac{1}{2}(\gamma - 1)M_2^2 = \frac{\frac{1}{2}(\gamma + 1)\left(1 + \left(\frac{\gamma+1}{2\gamma}\right)z\right)}{1 + z}, \tag{6.3.25}$$

$$1 + \gamma M_2^2 = \frac{(\gamma + 1)\left(1 + \frac{1}{2}z\right)}{1 + z},$$

and (6.3.16) becomes

$$\frac{u_2}{u_1} = \frac{1 + \left(\frac{\gamma-1}{2\gamma}\right)z}{1 + \left(\frac{\gamma+1}{2\gamma}\right)z}, \qquad \frac{p_2}{p_1} = 1 + z, \qquad \frac{\rho_2}{\rho_1} = \frac{1 + \left(\frac{\gamma+1}{2\gamma}\right)z}{1 + \left(\frac{\gamma-1}{2\gamma}\right)z},$$

$$\frac{T_2}{T_1} = \frac{(1 + z)\left(1 + \left(\frac{\gamma-1}{2\gamma}\right)z\right)}{1 + \left(\frac{\gamma+1}{2\gamma}\right)z}, \qquad \frac{c_2}{c_1} = \frac{(1 + z)^{\frac{1}{2}}\left(1 + \left(\frac{\gamma-1}{2\gamma}\right)z\right)^{\frac{1}{2}}}{\left(1 + \left(\frac{\gamma+1}{2\gamma}\right)z\right)^{\frac{1}{2}}}, \tag{6.3.26}$$

$$s_2 - s_1 = \frac{R}{\gamma - 1}\ln\left\{(1 + z)\left[\frac{1 + \left(\frac{\gamma-1}{2\gamma}\right)z}{1 + \left(\frac{\gamma+1}{2\gamma}\right)z}\right]^{\gamma}\right\}.$$

For weak shocks (i.e., $z \ll 1$), the first terms in the Taylor series give

$$M_2^2 \simeq 1 - \left(\frac{\gamma + 1}{2\gamma}\right)z,$$

$$1 + \frac{1}{2}(\gamma - 1)M_2^2 \simeq \frac{1}{2}(\gamma + 1)\left(1 - \left(\frac{\gamma - 1}{2\gamma}\right)z\right), \tag{6.3.27}$$

$$1 + \gamma M_2^2 \simeq (\gamma + 1)\left(1 - \frac{1}{2}z\right),$$

and

$$\frac{u_2}{u_1} \simeq 1 - \frac{z}{\gamma}, \qquad \frac{p_2}{p_1} = 1 + z, \qquad \frac{\rho_2}{\rho_1} \simeq 1 + \frac{z}{\gamma},$$

$$\frac{T_2}{T_1} \simeq 1 + \left(\frac{\gamma - 1}{\gamma}\right)z, \qquad \frac{c_2}{c_1} \simeq 1 + \left(\frac{\gamma - 1}{2\gamma}\right)z, \tag{6.3.28}$$

$$s_2 - s_1 \simeq R\left(\frac{\gamma + 1}{12\gamma^2}\right)z^3.$$

For strong shocks (i.e., $z \gg 1$), we have

$$M_1^2 \simeq \left(\frac{\gamma + 1}{2\gamma}\right)z, \qquad M_2^2 \simeq \frac{\gamma - 1}{2\gamma},$$

$$1 + \frac{1}{2}(\gamma - 1)M_2^2 \simeq \frac{(\gamma + 1)^2}{4\gamma}, \qquad 1 + \gamma M_2^2 \simeq \frac{1}{2}(\gamma + 1), \tag{6.3.29}$$

and

$$\frac{u_2}{u_1} \simeq \frac{\gamma - 1}{\gamma + 1}, \qquad \frac{p_2}{p_1} = z + 1, \qquad \frac{\rho_2}{\rho_1} \simeq \frac{\gamma + 1}{\gamma - 1},$$

$$\frac{T_2}{T_1} \simeq \left(\frac{\gamma - 1}{\gamma + 1}\right)z, \qquad \frac{c_2}{c_1} \simeq \left(\frac{\gamma - 1}{\gamma + 1}\right)^{\frac{1}{2}} z^{\frac{1}{2}}, \qquad s_2 - s_1 \simeq \frac{R}{\gamma - 1}\ln z. \tag{6.3.30}$$

Note from (6.3.29) and (6.3.30) that M_2 and u_2/u_1 are bounded away from zero, and ρ_2/ρ_1 is bounded from above. Thus a shock, however strong, has only a limited capacity to decrease the Mach number and flow speed, and thereby induce compression. In a diatomic gas, for which $\gamma = 7/5$, we have $(\gamma - 1)/(2\gamma) = 1/7$ and $(\gamma - 1)/(\gamma + 1) = 1/6$, so that

$$M_2^2 > \frac{1}{7}, \qquad \frac{u_2}{u_1} > \frac{1}{6}, \qquad \frac{\rho_2}{\rho_1} < 6. \tag{6.3.31}$$

Therefore in air, regarded as a diatomic gas with $\gamma = 7/5$, the strongest shock cannot induce more than a six-fold increase in density. These results for strong shocks are based on the assumption that the chemical structure of the gas does not change. In practice, at high enough incident Mach numbers the gas dissociates.

6.4 Oblique Shocks. The Shock Polar

Shocks that are not at right angles to the oncoming flow are said to be oblique and are universal in flows at high enough speeds, whether in the laboratory, or in engineering machinery, or around high-speed vehicles. At an oblique shock, the flow changes direction, so that there are three directions of interest, namely those of the upstream and downstream flows and that of the shock itself. An oblique shock may be attached or detached, as illustrated in Figure 6.4.1. The angle between the upstream flow and the shock is the shock angle ϕ, and the angle between the upstream and downstream flow is the deflection angle θ. The upstream Mach number is M_1, and the downstream Mach number is M_2.

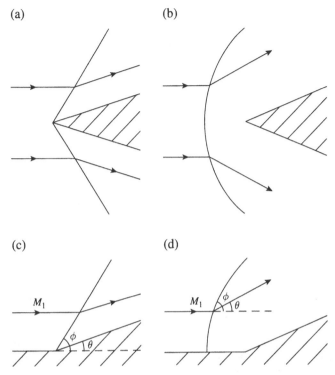

Fig. 6.4.1. Attached and detached shocks at a wedge and corner. Boundary layers are ignored. Upstream Mach number M_1, shock angle ϕ, deflection angle θ. (a) Attached, wedge; (b) detached, wedge; (c) attached, corner; (d) detached, corner.

We begin with a summary of factual information about oblique shocks. This information is rather complicated. We deal only with a polytropic gas, for example air that is not too hot, and we work in a frame of reference in which the shock is at rest. Then the upstream flow is necessarily supersonic (i.e., $M_1 > 1$), but the downstream flow may be either supersonic or subsonic, depending on the values of M_1 and ϕ. For an oblique shock that is part of a complicated flow pattern, the conditions determining whether the downstream flow is supersonic or subsonic are of great importance in practice, because they determine whether conditions further downstream have any upstream influence. Thus we shall attach importance to the transitional case $M_2 = 1$ in which the downstream flow is exactly sonic, and we shall give several formulae determining when this case occurs.

At a single value of M_1, the graph of θ against ϕ is as shown in Figure 6.4.2. The curve ABCD rises from $(\phi, \theta) = (\phi_{\min}(M_1), 0) = (\sin^{-1}(1/M_1), 0)$ at A

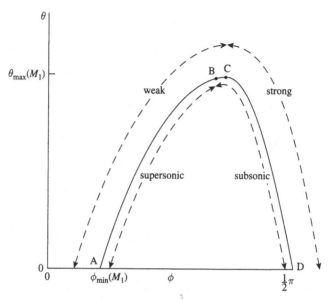

Fig. 6.4.2. Angles at an oblique shock in a polytropic gas. The curve ABCD gives deflection angle θ as a function of shock angle ϕ for given supersonic upstream flow at Mach number M_1. Arcs (AC, CD) correspond to (weak, strong) shocks; arcs (AB, BD) correspond to (supersonic, subsonic) downstream flow, that is, $M_2 > 1$ and $M_2 < 1$; A, at which $\phi = \phi_{\min}(M_1) = \sin^{-1}(1/M_1)$, corresponds to a Mach wave; B, at which ϕ is given by (6.4.10) or (6.4.11), corresponds to $M_2 = 1$; C, at which ϕ is given by (6.4.19), or by the vanishing of the numerator in (6.4.16), corresponds to the maximum deflection $\theta_{\max}(M_1)$; D, at which $\phi = \frac{1}{2}\pi$, corresponds to a normal shock; and values of θ are given by (6.4.15), the equation of ABCD.

to a maximum height $\theta = \theta_{\max}(M_1)$ at C and then falls to $(\phi, \theta) = (\frac{1}{2}\pi, 0)$ at D. Therefore the shock angle is confined to the interval $\phi_{\min}(M_1) \leq \phi \leq \frac{1}{2}\pi$ and the deflection angle is confined to $0 \leq \theta \leq \theta_{\max}(M_1)$. For a given Mach number $M_1 \geq 1$, the angle $\sin^{-1}(1/M_1)$ occurs repeatedly in what follows and is the Mach angle. As there are two values of ϕ for given θ, but only one value of θ for given ϕ, it is convenient to express formulae in terms of ϕ not θ. Later in this section we give complete formulae for all curves and special points in our graphs.

At point A, where the shock angle equals the Mach angle, the shock has vanishing strength and is described by the equations of acoustics. It is therefore a Mach wave rather than a shock wave and is a characteristic (i.e., a Mach surface or Mach line). Photographs of supersonic flow next to a rough surface make clearly visible a fine grating of nearly parallel Mach lines.

On AC a shock is said to be weak, and on CD it is said to be strong. At C, the deflection $\theta = \theta_{\max}(M_1)$ is the maximum possible by an oblique shock alone, so that the larger total deflection produced by a wide-angle wedge or a strongly concave wall occurs by means of a detached shock and curved streamlines, as shown in Figure 6.4.1b,d. Thus if θ is given, and $\theta > \theta_{max}(M_1)$, then no ϕ exists satisfying the shock conditions, whereas if $\theta < \theta_{max}(M_1)$ there are two such ϕ, and the smaller ϕ corresponds to the weaker shock. In Chapter 11, we discuss which shock can be expected to occur in practice; it is usually the weaker. As we have seen, when $\theta \to 0$ the weak shock tends to a Mach wave, for which $\phi = \sin^{-1}(1/M_1)$, corresponding to point A, and the strong shock tends to a normal shock, for which $\phi = \frac{1}{2}\pi$, corresponding to point D.

On AB, the flow downstream of the shock is supersonic, and on BD it is subsonic. Point B is always to the left of C and very close to it: The values of ϕ never differ by more than 4.5°, nor do the values of θ differ by more than 0.5°. Therefore the flow through a weak shock nearly always keeps the speed supersonic, and that through a strong shock always makes the speed subsonic. Since the range of deflection angles for which a weak shock has subsonic flow behind it is less than 0.5°, an observed stationary oblique shock with subsonic flow behind it is nearly always a strong shock. A normal shock, whatever its strength, is always a strong shock according to our definition, because it corresponds to point D in Figure 6.4.2, and so it always has subsonic flow behind it, as we found in (6.2.11).

As M_1 varies, the curve in Figure 6.4.2 varies as in Figure 6.4.3. When M_1 decreases to 1, so the Mach angle $\sin^{-1}(1/M_1)$ increases to $\frac{1}{2}\pi$, the range of possible shock angles ϕ narrows to zero, and the maximum deflection angle $\theta_{max}(M_1)$ decreases to zero. Thus in the limit $M_1 \to 1$, the curve shrinks to the point $(\phi, \theta) = (\frac{1}{2}\pi, 0)$. When M_1 increases to infinity, so the Mach angle decreases to zero, the range of shock angles widens to $\frac{1}{2}\pi$, and the maximum deflection angle increases to a value we shall call $\hat{\theta}$, corresponding to a shock angle $\hat{\phi}$. The curve expands to approach a smooth limiting curve of the same general shape as for large finite M_1. Recall that we are discussing a polytropic gas with ratio of specific heats γ. We shall see below that $\theta_{max}(M_1) \to \sin^{-1}(1/\gamma)$ as $M_1 \to \infty$, and hence $(\hat{\phi}, \hat{\theta}) = (\sin^{-1}(\{(\gamma+1)/(2\gamma)\}^{\frac{1}{2}}), \sin^{-1}(1/\gamma))$. For a diatomic gas, we have $\gamma = 7/5$, so that $(\hat{\phi}, \hat{\theta}) = (\sin^{-1}((6/7)^{\frac{1}{2}}), \sin^{-1}(5/7)) = (67.8°, 45.6°)$. That is, in a diatomic gas the maximum possible flow deflection at an oblique shock is 45.6°, and the corresponding shock angle is 67.8°.

The curve marked $\theta = \theta_{max}(M_1)$ in Figure 6.4.3 is the locus, as M_1 increases from 1 to infinity, of point C in Figure 6.4.2. It starts at $(\phi, \theta) = (\frac{1}{2}\pi, 0)$ with a horizontal tangent, extends upwards and left to a minimum value of ϕ, and continues upwards and right to its highest point $(\hat{\phi}, \hat{\theta})$. The curve separates the (ϕ, θ) plane into a left-hand region of weak shocks and a right-hand region of strong shocks. The function $\theta = \theta_{max}(M_1)$, which may equally be written

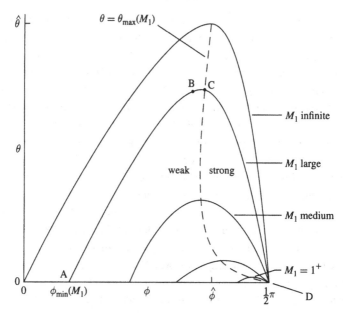

Fig. 6.4.3. Extension of Figure 6.4.2 to include dependence of deflection angle θ on upstream Mach number M_1. The maximum possible deflection angle lies on the limiting curve as $M_1 \to \infty$ and is denoted $\hat{\theta}$; the corresponding shock angle is $\hat{\phi}$. For a diatomic gas, with $\gamma = 1.4$, these angles are $\hat{\theta} = \sin^{-1}(\frac{5}{7}) = 45.6°$ and $\hat{\phi} = \sin^{-1}((\frac{6}{7})^{\frac{1}{2}}) = 67.8°$.

$M_1 = M_{1,\min}(\theta)$, is plotted in the (M_1, θ) plane in Figure 6.4.4a. The lower right region corresponds to an attached shock, at two possible angles, and the upper left region, with θ reinterpreted as the wall angle θ shown in Figure 6.4.1c, corresponds to a detached shock. As $M_1 \to \infty$, the curve asymptotes to $\theta = \hat{\theta}$. Thus in a diatomic gas, if $\theta > 45.6°$ then a shock never attaches for any M_1, but if $\theta < 45.6°$ then a shock can be made to attach by taking M_1 large enough (i.e., $M_1 > M_{1,\min}(\theta)$).

The information presented so far is replotted in the (M_1, ϕ) plane in Figure 6.4.4b. This is a most convenient plane in the theory of oblique shocks, and the reader may find it helpful to refer regularly to the figure in interpreting the algebraic formulae in the rest of the chapter.

To obtain the algebraic formulae for an oblique shock, we use the notation in Figure 6.4.5 and express all quantities in terms of ϕ and $M_1 = u_1/c_1$. We have $u_2 = w_2 \sin(\phi - \theta)$, $v_2 = w_2 \cos(\phi - \theta)$, and

$$\frac{u_2}{v_2} = \tan(\phi - \theta). \qquad (6.4.1)$$

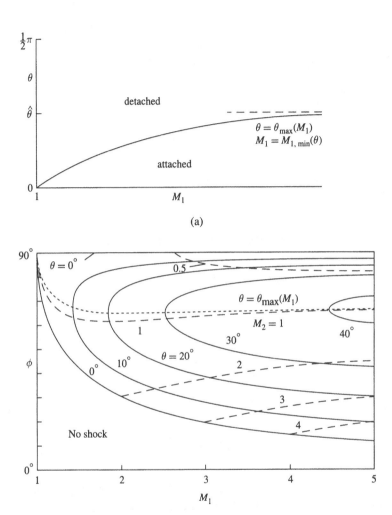

(a)

(b)

Fig. 6.4.4. (a) Regions in the (M_1, θ) plane corresponding to attached and detached shocks in a polytropic gas. For a diatomic gas, $\hat{\theta} = 45.6°$. (b) The (M_1, ϕ) plane. ——, contours $\theta = 0°, 10°, 20°, 30°, 40°$; – – –, contours $M_2 = 0.5, 1, 2, 3, 4$; - - - -, points (M_1, ϕ) of maximum θ at fixed M_1 (i.e., points where the θ-contours are vertical). The curved part of the contour $\theta = 0°$ corresponds to a Mach wave and has equation $\phi = \sin^{-1}(1/M_1)$; the straight part, forming the line $\phi = 90°$, corresponds to a normal shock. Thus $M_2 = M_1$ on the curved part, and no shock is possible below it. The contour $M_2 = 1$ corresponds to exactly sonic downstream speed. All curves are for $\gamma = 1.4$.

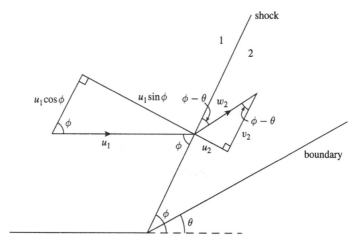

Fig. 6.4.5. Velocity components at an oblique shock.

Continuity of the tangential component of velocity gives

$$v_2 = u_1 \cos \phi. \tag{6.4.2}$$

Therefore the downstream Mach number M_2 satisfies

$$
\begin{aligned}
M_2^2 &= \left(\frac{w_2}{c_2}\right)^2 = \left(\frac{u_2}{c_2}\right)^2 + \left(\frac{v_2}{c_2}\right)^2 \\
&= \left(\frac{u_2}{c_2}\right)^2 + \left(\frac{u_1 \cos \phi}{c_2}\right)^2 \\
&= \left(\frac{u_2}{c_2}\right)^2 + \left(\frac{c_1}{c_2}\right)^2 M_1^2 \cos^2 \phi.
\end{aligned}
\tag{6.4.3}
$$

The equations for the normal components of velocities are obtained from the normal shock equations in Section 6.3 by making the substitutions

$$u_1 \mapsto u_1 \sin \phi, \qquad M_1 \mapsto M_1 \sin \phi, \qquad M_2 \mapsto \frac{u_2}{c_2}. \tag{6.4.4}$$

The resulting equations determine all the required quantities at an oblique shock, except that they do not contain M_2, which is found at the end of the calculation from (6.4.3). Thus (6.3.16) gives

$$\frac{u_2}{u_1} = \frac{1 + \frac{1}{2}(\gamma - 1)M_1^2 \sin^2 \phi}{\frac{1}{2}(\gamma + 1)M_1^2 \sin \phi}, \tag{6.4.5a}$$

$$\frac{p_2}{p_1} = 1 + \frac{2\gamma}{\gamma + 1}(M_1^2 \sin^2 \phi - 1) = \frac{\gamma M_1^2 \sin^2 \phi - \frac{1}{2}(\gamma - 1)}{\frac{1}{2}(\gamma + 1)}, \tag{6.4.5b}$$

$$\frac{\rho_2}{\rho_1} = \frac{\frac{1}{2}(\gamma + 1)M_1^2 \sin^2 \phi}{1 + \frac{1}{2}(\gamma - 1)M_1^2 \sin^2 \phi}, \tag{6.4.5c}$$

$$\frac{T_2}{T_1} = \frac{\left(1 + \frac{1}{2}(\gamma - 1)M_1^2 \sin^2 \phi\right)\left(\gamma M_1^2 \sin^2 \phi - \frac{1}{2}(\gamma - 1)\right)}{\frac{1}{4}(\gamma + 1)^2 M_1^2 \sin^2 \phi}, \tag{6.4.5d}$$

$$\frac{c_2}{c_1} = \frac{\left(1 + \frac{1}{2}(\gamma - 1)M_1^2 \sin^2 \phi\right)^{\frac{1}{2}}\left(\gamma M_1^2 \sin^2 \phi - \frac{1}{2}(\gamma - 1)\right)^{\frac{1}{2}}}{\frac{1}{2}(\gamma + 1)M_1 \sin \phi}, \tag{6.4.5e}$$

$$s_2 - s_1 = \frac{R}{\gamma - 1} \ln \left\{ \frac{\gamma M_1^2 \sin^2 \phi - \frac{1}{2}(\gamma - 1)}{\frac{1}{2}(\gamma + 1)} \left(\frac{1 + \frac{1}{2}(\gamma - 1)M_1^2 \sin^2 \phi}{\frac{1}{2}(\gamma + 1)M_1^2 \sin^2 \phi} \right)^{\gamma} \right\}, \tag{6.4.5f}$$

and (6.3.6) gives

$$\left(\frac{u_2}{c_2}\right)^2 = \frac{1 + \frac{1}{2}(\gamma - 1)M_1^2 \sin^2 \phi}{\gamma M_1^2 \sin^2 \phi - \frac{1}{2}(\gamma - 1)}. \tag{6.4.6}$$

Therefore (6.4.3) is

$$M_2^2 = \frac{\left(1 + \frac{1}{2}(\gamma - 1)M_1^2 \sin^2 \phi\right)^2 + \left(\frac{1}{2}(\gamma + 1)M_1^2 \sin \phi \cos \phi\right)^2}{\left(1 + \frac{1}{2}(\gamma - 1)M_1^2 \sin^2 \phi\right)\left(\gamma M_1^2 \sin^2 \phi - \frac{1}{2}(\gamma - 1)\right)}. \tag{6.4.7}$$

Collecting powers of M_1 in the numerator, we obtain

$$M_2^2 = \frac{1 + (\gamma - 1)M_1^2 \sin^2 \phi + M_1^4\left(\frac{1}{4}(\gamma + 1)^2 - \gamma \sin^2 \phi\right)\sin^2 \phi}{\left(1 + \frac{1}{2}(\gamma - 1)M_1^2 \sin^2 \phi\right)\left(\gamma M_1^2 \sin^2 \phi - \frac{1}{2}(\gamma - 1)\right)}. \tag{6.4.8}$$

It may be checked that $\phi = \sin^{-1}(1/M_1)$ gives $M_2 = M_1$, as expected for a Mach wave, and that $\phi = \frac{1}{2}\pi$ gives $M_2^2 = \{1 + \frac{1}{2}(\gamma - 1)M_1^2\}/\{\gamma M_1^2 - \frac{1}{2}(\gamma - 1)\}$, that is, (6.3.6) for a normal shock. The limit $M_1 \to \infty$ gives

$$M_2^2 \to \frac{\frac{1}{4}(\gamma + 1)^2 - \gamma \sin^2 \phi}{\frac{1}{2}\gamma(\gamma - 1)\sin^2 \phi}. \tag{6.4.9}$$

On putting $M_2 = 1$ in (6.4.8) we obtain

$$\gamma(M_1 \sin \phi)^4 - \left(\frac{1}{2}(\gamma - 3) + \frac{1}{2}(\gamma + 1)M_1^2\right)(M_1 \sin \phi)^2 - 1 = 0. \tag{6.4.10}$$

This may be regarded as a quadratic equation in $\sin^2 \phi$, with positive root

$$\sin^2 \phi = \frac{\frac{1}{2}(\gamma - 3) + \frac{1}{2}(\gamma + 1)M_1^2 + (\gamma + 1)^{\frac{1}{2}}\left\{\frac{1}{4}(\gamma + 9) + \frac{1}{2}(\gamma - 3)M_1^2 + \frac{1}{4}(\gamma + 1)M_1^4\right\}^{\frac{1}{2}}}{2\gamma M_1^2}. \tag{6.4.11}$$

Thus Equations (6.4.10) and (6.4.11) give the curve in the (M_1, ϕ) plane corresponding to exactly sonic downstream velocity. That is, they give ϕ as a function of M_1 for point B in Figure 6.4.2 and for the contour $M_2 = 1$ in Figure 6.4.4b. On this contour, (6.4.11) shows that $\phi \to \sin^{-1}(\{(\gamma + 1)/(2\gamma)\}^{\frac{1}{2}})$ as $M_1 \to \infty$ and that for $M_1 \to 1$ the contour is approximated by

$$\phi \simeq \frac{\pi}{2} - \left\{\frac{1}{2}(M_1^2 - 1)\right\}^{\frac{1}{2}}. \tag{6.4.12}$$

A contour plot comparable to that for M_2 in Figure 6.4.4b may readily be constructed for any of the quantities in (6.4.5)–(6.4.6).

The deflection angle θ is given by (6.4.1). Substituting for v_2 from (6.4.2) gives

$$\tan(\phi - \theta) = \frac{u_2}{u_1 \cos \phi}. \tag{6.4.13}$$

Therefore

$$\frac{\tan \phi - \tan \theta}{1 + \tan \phi \tan \theta} = \frac{1 + \frac{1}{2}(\gamma - 1)M_1^2 \sin^2 \phi}{\frac{1}{2}(\gamma + 1)M_1^2 \sin \phi \cos \phi}. \tag{6.4.14}$$

This equation, linear in $\tan \theta$, has solution

$$\tan \theta = \frac{\left(M_1^2 \sin^2 \phi - 1\right)\cot \phi}{1 + \left(\frac{1}{2}(\gamma + 1) - \sin^2 \phi\right)M_1^2}. \tag{6.4.15}$$

Equation (6.4.15), giving the deflection angle θ explicitly as a function of the shock angle ϕ, is most useful. It gives Figures 6.4.2–6.4.3, and in conjunction with Equations (6.4.5)–(6.4.8) it gives the results quoted in the first half of this section. Substitution of the value of ϕ from (6.4.11) into (6.4.15) gives the deflection angle θ producing exactly sonic downstream velocity when the upstream Mach number is M_1; that is, it gives θ as a function of M_1 on the contour $M_2 = 1$ in Figure 6.4.4b. Differentiation of (6.4.15) at fixed M_1 gives

$$\frac{d\theta}{d\phi} = \frac{\gamma(M_1 \sin \phi)^4 + 2\left(1 - \frac{1}{4}(\gamma + 1)M_1^2\right)(M_1 \sin \phi)^2 - \left(1 + \frac{1}{2}(\gamma + 1)M_1^2\right)}{\gamma(M_1 \sin \phi)^4 - \left((\gamma - 1) + \frac{1}{4}(\gamma + 1)^2 M_1^2\right)(M_1 \sin \phi)^2 - 1}. \tag{6.4.16}$$

Thus at the Mach angle $\phi = \sin^{-1}(1/M_1)$ we obtain

$$\frac{d\theta}{d\phi} = \frac{4\left(M_1^2 - 1\right)}{(\gamma + 1)M_1^2}, \tag{6.4.17}$$

and at the normal shock angle $\phi = \frac{1}{2}\pi$ we obtain, on cancelling a factor $1 + \frac{1}{2}(\gamma - 1)M_1^2$,

$$\frac{d\theta}{d\phi} = \frac{-(M_1^2 - 1)}{1 + \frac{1}{2}(\gamma - 1)M_1^2}. \tag{6.4.18}$$

The numerator of (6.4.16) is zero when

$$\sin^2 \phi = \frac{-1 + \frac{1}{4}(\gamma + 1)M_1^2 + (\gamma + 1)^{\frac{1}{2}}\left\{1 + \frac{1}{2}(\gamma - 1)M_1^2 + \frac{1}{16}(\gamma + 1)M_1^4\right\}^{\frac{1}{2}}}{\gamma M_1^2}. \tag{6.4.19}$$

This gives the value of ϕ corresponding to maximum deflection. Thus if ϕ from (6.4.19) is substituted into (6.4.15), we obtain the function $\theta_{max}(M_1)$ indicated in Figures 6.4.2–6.4.4.

The deflection as $M_1 \to \infty$ is of great interest. Then (6.4.15) becomes

$$\tan \theta = \frac{\cos \phi \sin \phi}{\frac{1}{2}(\gamma + 1) - \sin^2 \phi} = \frac{\sin 2\phi}{\gamma + \cos 2\phi}. \tag{6.4.20}$$

Hence

$$\frac{d\theta}{d\phi} = \frac{\frac{1}{2}(\gamma + 1) - \gamma \sin^2 \phi}{\frac{1}{4}(\gamma + 1)^2 - \gamma \sin^2 \phi} = \frac{1 + \gamma \cos 2\phi}{\frac{1}{2}(\gamma^2 + 1) + \gamma \cos 2\phi}. \tag{6.4.21}$$

Therefore at the Mach angle, now $\phi \simeq 1/M_1$, we have $d\theta/d\phi = 2/(\gamma + 1)$, and at the normal shock angle $\phi = \frac{1}{2}\pi$ we have $d\theta/d\phi = -2/(\gamma - 1)$. The numerators in (6.4.21) are zero when $\sin \phi = \{(\gamma + 1)/(2\gamma)\}^{\frac{1}{2}}$. This gives the value $\hat{\phi} = \sin^{-1}(\{(\gamma + 1)/(2\gamma)\}^{\frac{1}{2}})$ corresponding to the highest point on the highest curve in Figure 6.4.3. Thus $\cot \hat{\phi} = \{(\gamma - 1)/(\gamma + 1)\}^{\frac{1}{2}}$, and from (6.4.15) the maximum deflection $\hat{\theta}$ is given by $\tan \hat{\theta} = 1/(\gamma^2 - 1)^{\frac{1}{2}}$ (i.e., $\sin \hat{\theta} = 1/\gamma$). These results for $M_1 \to \infty$, obtained from (6.4.20), may also be obtained by taking the limit $M_1 \to \infty$ in Equations (6.4.16)–(6.4.19), with one exception. Near the origin in Figure 6.4.3, the pattern of curves is similar to that near a saddle point, so that the slope of the limiting curve is different from the limit of the neighbouring slopes. Thus (6.4.17) tends to $4/(\gamma + 1)$ as $M_1 \to \infty$, whereas the slope of the limiting curve at the origin is in fact $2/(\gamma + 1)$, as we have seen. Although this discontinuity in slope is easy to overlook, a close inspection of published graphs shows that it is really present, confirming that the creators of the graphs have done their job accurately.

Of equal interest is the limit $M_1 \to 1$. Then the shock angle ϕ for exactly sonic downstream velocity is given by (6.4.12), and the corresponding deflection angle is

$$\theta \simeq \frac{2}{\gamma+1}\left\{\frac{1}{2}\left(M_1^2-1\right)\right\}^{\frac{3}{2}}. \tag{6.4.22}$$

From (6.4.19), the maximum deflection occurs for

$$\phi \simeq \frac{\pi}{2} - \left\{\frac{1}{3}\left(M_1^2-1\right)\right\}^{\frac{1}{2}}, \tag{6.4.23}$$

and from (6.4.15) it is

$$\theta_{\max} \simeq \frac{4}{\gamma+1}\left\{\frac{1}{3}\left(M_1^2-1\right)\right\}^{\frac{3}{2}}. \tag{6.4.24}$$

To obtain Prandtl's relation for an oblique shock, we start with the equation corresponding to (6.3.19a), that is,

$$\frac{c_1^2}{\gamma-1} + \frac{(u_1\sin\phi)^2}{2} + \frac{(u_1\cos\phi)^2}{2} = \frac{c_2^2}{\gamma-1} + \frac{u_2^2}{2} + \frac{v_2^2}{2} = \frac{1}{2}\left(\frac{\gamma+1}{\gamma-1}\right)\hat{c}^2. \tag{6.4.25}$$

Since $v_2 = u_1\cos\phi$, by (6.4.2), this is

$$\frac{c_1^2}{\gamma-1} + \frac{(u_1\sin\phi)^2}{2} = \frac{c_2^2}{\gamma-1} + \frac{u_2^2}{2} = \frac{1}{2}\left(\frac{\gamma+1}{\gamma-1}\right)\hat{c}^2 - \frac{v_2^2}{2}. \tag{6.4.26}$$

The algebra is now similar to that in (6.3.18)–(6.3.21) and leads to

$$u_1 u_2 \sin\phi = \hat{c}^2 - \left(\frac{\gamma-1}{\gamma+1}\right)v_2^2. \tag{6.4.27}$$

This is Prandtl's relation for an oblique shock with tangential component of velocity v_2.

Downstream of an oblique shock, let the velocity components parallel and perpendicular to the upstream flow be (u, v), as shown in Figure 6.4.6a, so that

$$\begin{aligned}u &= u_2\sin\phi + v_2\cos\phi, \\ v &= -u_2\cos\phi + v_2\sin\phi.\end{aligned} \tag{6.4.28}$$

The (u, v) plane is the hodograph plane, and the curve representing all possible downstream velocities for given upstream conditions is the shock polar. As M_1 varies from 1 to infinity, the shock polar varies as in Figure 6.4.6b. Given the deflection angle θ, the relation $\tan\theta = v/u$ implies that (u, v) is at an

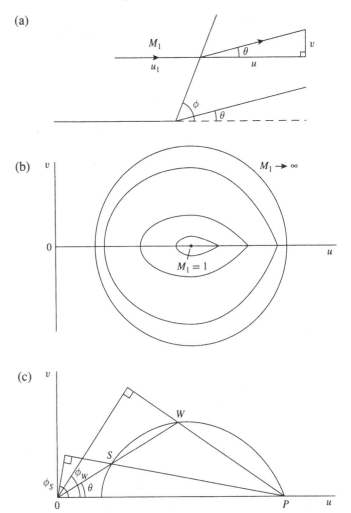

Fig. 6.4.6. The shock polar for a polytropic gas. (a) Downstream components of velocity, (u, v). (b) Shock polars for different upstream Mach numbers M_1. (c) Construction of strong shock angle ϕ_S and weak shock angle ϕ_W, given the deflection angle θ.

intersection of the polar with a line through the origin at an angle θ to the u axis. Two such intersection points, marked S and W to indicate the strong and weak shocks, are shown in Figure 6.4.6c. From the right-most point P on the polar, we may construct lines PS and PW and their perpendiculars to the origin. The angles between these perpendiculars and the u axis are the shock angles, say ϕ_S and ϕ_W, corresponding to θ. We saw that ϕ_W approaches the Mach angle $\sin^{-1}(1/M_1)$ as $\theta \to 0$. Therefore at P the tangent to the polar is inclined to the u axis at the complement of the Mach angle.

The equation of the shock polar follows from (6.4.28) and a selection of (6.4.1)–(6.4.27). Using the critical speed $\hat{c}$ defined by (6.4.25), we may define normalised velocity components $\hat{u} = u/\hat{c}$, $\hat{v} = v/\hat{c}$, and $\hat{u}_1 = u_1/\hat{c}$, to obtain

$$\hat{v}^2 = \frac{(\hat{u}_1 - \hat{u})^2(\hat{u}_1\hat{u} - 1)}{1 - \hat{u}_1\hat{u} + \left(\frac{2}{\gamma+1}\right)\hat{u}_1^2}. \tag{6.4.29}$$

Therefore the polar is a strophoid (i.e., a folium of Descartes). The remarks in the last paragraph, and many of the results earlier in this section, may be obtained as consequences of (6.4.29). In the limit $M_1 \to \infty$ (e.g., $u_1 \to \infty$ at fixed c_1) Equation (6.4.25) gives $\hat{u}_1 = ((\gamma + 1)/(\gamma - 1))^{\frac{1}{2}}$, so that the denominator of (6.4.29) is $\hat{u}_1(\hat{u}_1 - \hat{u})$. Then the polar becomes the circle

$$\hat{v}^2 = (\hat{u}_1 - \hat{u})(\hat{u} - \hat{u}_1^{-1}). \tag{6.4.30}$$

The limiting form of the polar as $M_1 \to 1$ is also easy to obtain and shows that the scale of v, as measured by the vertical extent of the small eyes in Figure 6.4.6b, is then of order $(M_1^2 - 1)^{3/2}$.

Shock polars are often presented in terms of the variables p_2/p_1 and θ (i.e., pressure ratio and deflection). The resulting pressure–deflection polars are especially convenient in analysing shock reflections, for these often produce vortex sheets in directions that must be determined, and the boundary conditions between different regions of uniform flow separated by a vortex sheet are continuity of pressure and flow direction.

Pressure–deflection polars are shown for a single M_1 in Figure 6.4.7 and for several M_1 in Figure 6.4.8. They would ideally be drawn with the θ axis vertical, for comparison with other diagrams, but we shall follow the standard practice of drawing them with the θ axis horizontal. To plot them, (6.4.5b) may be used to express $\sin\phi$ in terms of p_2/p_1 and M_1; then substitution into (6.4.15) gives a formula, cumbersome but explicit, for $\theta(p_2/p_1, M_1)$. Alternatively, we may use ϕ as a parameter, since (6.4.5b) and (6.4.15) give p_2/p_1 and θ as rather simple functions of M_1 and ϕ. It is convenient to allow ϕ to vary in the range $\sin^{-1}(1/M_1) \le \phi \le \pi - \sin^{-1}(1/M_1)$, so that in Figure 6.4.1c, for example, a shock angle $\phi > \frac{1}{2}\pi$ corresponds to a wall at the top of the figure and $\theta < 0$. The transition through $\phi = \frac{1}{2}\pi$ is continuous, and all formulae have an obvious symmetry or antisymmetry in ϕ about $\frac{1}{2}\pi$. In Figure 6.4.8, $\phi = \sin^{-1}(1/M_1)$ corresponds to the point $(\theta, p_2/p_1) = (0, 1)$, representing a Mach wave; $\phi = \frac{1}{2}\pi$ corresponds to $(\theta, p_2/p_1) = (0, 1 + (2\gamma/(\gamma + 1))(M_1^2 - 1))$, representing a normal shock; and $\phi = \pi - \sin^{-1}(1/M_1)$ corresponds to $(\theta, p_2/p_1) = (0, 1)$, representing a Mach wave of the other family. Thus $(0, 1)$ is really two separate points, the two ends of the polar. Mathematically, the polar may be continued into the half-plane $p_2/p_1 < 1$, but the resulting points correspond to rarefaction

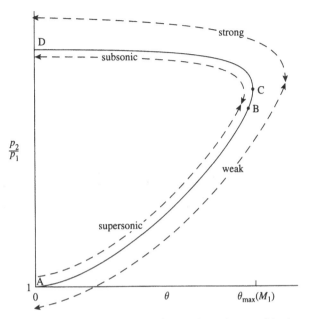

Fig. 6.4.7. Pressure–deflection shock polar for a polytropic gas with given upstream Mach number M_1, for shock angles ϕ in the range $\sin^{-1}(1/M_1) \leq \phi \leq \frac{1}{2}\pi$. If the θ axis is drawn vertically, and the p_2/p_1 axis is drawn horizontally and rescaled to ϕ at fixed M_1 in accordance with the formula $p_2/p_1 = 1 + (2\gamma/(\gamma+1))(M_1^2 \sin^2 \phi - 1)$, the curve becomes that in Figure 6.4.2, and the lettering corresponds. Thus $\phi = \sin^{-1}(1/M_1)$ at A corresponds to a Mach wave, and $\phi = \frac{1}{2}\pi$ at D corresponds to a normal shock. The function $\theta_{\max}(M_1)$ is the same as that shown in Figure 6.4.4a.

shocks, which cannot occur in a polytropic gas because of the entropy condition (6.1.1d). The point corresponding to exactly sonic downstream velocity (i.e., $M_2 = 1$) is obtained by substituting the value of ϕ from (6.4.11) into (6.4.15) and (6.4.5b).

The derivative $d\theta/d\phi$ on a pressure–deflection polar is given by (6.4.16), and we also have

$$\frac{1}{p_1}\frac{dp_2}{d\phi} = \frac{4\gamma}{\gamma+1}M_1^2 \sin\phi \cos\phi. \tag{6.4.31}$$

Division of the latter expression by the former shows that the slope of the polar is given parametrically by

$$\frac{1}{p_1}\frac{dp_2}{d\theta}$$
$$= \frac{(4\gamma/(\gamma+1))\{(\gamma M_1 \sin\phi)^4 - ((\gamma-1) + \frac{1}{4}(\gamma+1)M_1^2)(M_1 \sin\phi)^2 - 1\}M_1^2 \sin\phi \cos\phi}{\gamma(M_1 \sin\phi)^4 + 2(1 - \frac{1}{4}(\gamma+1)M_1^2)(M_1 \sin\phi)^2 - (1 + \frac{1}{2}(\gamma+1)M_1^2)}.$$
$$\tag{6.4.32}$$

Therefore three sets of values of $(\phi, d\theta/d\phi, (1/p_1)dp_2/d\phi, (1/p_1)dp_2/d\theta)$ on

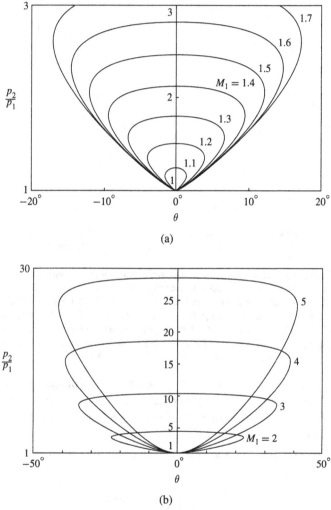

(a)

(b)

Fig. 6.4.8. Pressure–deflection shock polars for $\gamma = 1.4$ and different M_1, for shock angles ϕ in the range $\sin^{-1}(1/M_1) \le \phi \le \pi - \sin^{-1}(1/M_1)$. The right-most point on each polar is at $\theta = \theta_{max}(M_1)$, plotted in Figure 6.4.4. As $M_1 \to \infty$, $\theta_{max}(M_1) \to \hat{\theta} = 45.6°$, and the top of the polar tends to infinity. As $M_1 \to 1$ the polar shrinks and becomes much narrower than tall, before finally shrinking to the point $(\theta, p_2/p_1) = (0, 1)$. Thus in both limits the height of the polar is much greater than its width. (a) $M_1 = 1.1$, $1.2, \dots, 1.7$; (b) $M_1 = 2, 3, 4, 5$.

the polar are

$$
\left(\sin^{-1}\left(\frac{1}{M_1}\right), \ \frac{4(M_1^2-1)}{(\gamma+1)M_1^2}, \ \frac{4\gamma}{\gamma+1}(M_1^2-1)^{\frac{1}{2}}, \ \frac{\gamma M_1^2}{(M_1^2-1)^{\frac{1}{2}}} \right),
$$

$$
\left(\frac{\pi}{2}, \ \frac{-(M_1^2-1)}{1+\frac{1}{2}(\gamma-1)M_1^2}, \ 0, \ 0 \right), \tag{6.4.33}
$$

$$
\left(\pi-\sin^{-1}\left(\frac{1}{M_1}\right), \ \frac{4(M_1^2-1)}{(\gamma+1)M_1^2}, \ \frac{-4\gamma}{\gamma+1}(M_1^2-1)^{\frac{1}{2}}, \ \frac{-\gamma M_1^2}{(M_1^2-1)^{\frac{1}{2}}} \right).
$$

The point of maximum deflection $\theta_{\max}(M_1)$ is obtained by substituting the value of ϕ from (6.4.19) into (6.4.15) and (6.4.5b). The supplementary angle to ϕ gives the deflection $-\theta_{\max}(M_1)$, at the same value of p_2/p_1.

When M_1 is very large, the height of the polar may be approximated by $(2\gamma/(\gamma+1))M_1^2$, and its half-width by $\sin^{-1}(1/\gamma)$. Its parametric equation is then approximately

$$
\tan\theta = \frac{\sin\phi\cos\phi}{\frac{1}{2}(\gamma+1)-\sin^2\phi}, \quad \frac{1}{M_1^2}\frac{p_2}{p_1} = \frac{2\gamma}{\gamma+1}\sin^2\phi. \tag{6.4.34}
$$

The sets of values (6.4.33) determining the slopes become

$$
\left(\frac{1}{M_1}, \ \frac{4}{\gamma+1}, \ \frac{4\gamma M_1}{\gamma+1}, \ \gamma M_1 \right),
$$

$$
\left(\frac{\pi}{2}, \ \frac{-2}{\gamma-1}, \ 0, \ 0 \right), \tag{6.4.35}
$$

$$
\left(\pi-\frac{1}{M_1}, \ \frac{4}{\gamma+1}, \ \frac{-4\gamma M_1}{\gamma+1}, \ -\gamma M_1 \right).
$$

To leading order in M_1, the points of exactly sonic downstream velocity and of maximum deflection are now coincident, and for positive θ they are at

$$
\phi \simeq \sin^{-1}\left(\left(\frac{\gamma+1}{2\gamma}\right)^{\frac{1}{3}}\right), \quad \theta \simeq \sin^{-1}\left(\frac{1}{\gamma}\right), \quad \frac{p_2}{p_1} \simeq M_1^2. \tag{6.4.36}
$$

To leading order, the lower part of the polar, below the point of maximum deflection, has height M_1^2, and the upper part, above this point, has height $((\gamma-1)/(\gamma+1))M_1^2$. Thus for a diatomic gas, with $\gamma=7/5$, the lower part is six times as tall as the upper part.

When $M_1 \to 1$, the half-width of the polar is approximately $(4/(\gamma+1))\{\frac{1}{3}(M_1^2-1)\}^{3/2}$, and its height is still $(2\gamma/(\gamma+1))(M_1^2-1)$. Thus it shrinks to the point $(\theta, p_2/p_1) = (0, 1)$ and is much narrower than tall, by a factor of order $(M_1^2 - 1)^{\frac{1}{2}}$. The sets of values (6.4.33) determining the slopes become

$$\left(\frac{\pi}{2} - (M_1^2 - 1)^{\frac{1}{2}}, \quad \frac{4}{\gamma+1}(M_1^2 - 1), \quad \frac{4\gamma}{\gamma+1}(M_1^2 - 1)^{\frac{1}{2}}, \quad \frac{\gamma}{(M_1^2 - 1)^{\frac{1}{2}}} \right),$$

$$\left(\frac{\pi}{2}, \quad \frac{-2}{\gamma+1}(M_1^2 - 1), \quad 0, \quad 0 \right), \qquad (6.4.37)$$

$$\left(\frac{\pi}{2} + (M_1^2 - 1)^{\frac{1}{2}}, \quad \frac{4}{\gamma+1}(M_1^2 - 1), \quad \frac{-4\gamma}{\gamma+1}(M_1^2 - 1)^{\frac{1}{2}}, \quad \frac{-\gamma}{(M_1^2 - 1)^{\frac{1}{2}}} \right).$$

For positive θ, the point of exactly sonic downstream velocity is

$$\phi \simeq \frac{\pi}{2} - \left\{ \frac{1}{2}(M_1^2 - 1) \right\}^{\frac{1}{2}}, \qquad \theta \simeq \frac{2}{\gamma+1} \left\{ \frac{1}{2}(M_1^2 - 1) \right\}^{\frac{3}{2}},$$

$$\frac{p_2}{p_1} \simeq 1 + \frac{\gamma}{\gamma+1}(M_1^2 - 1), \qquad (6.4.38)$$

and the point of maximum deflection is

$$\phi \simeq \frac{\pi}{2} - \left\{ \frac{1}{3}(M_1^2 - 1) \right\}^{\frac{1}{2}}, \qquad \theta \simeq \frac{4}{\gamma+1} \left\{ \frac{1}{3}(M_1^2 - 1) \right\}^{\frac{3}{2}},$$

$$\frac{p_2}{p_1} \simeq 1 + \frac{4}{3} \left(\frac{\gamma}{\gamma+1} \right)(M_1^2 - 1). \qquad (6.4.39)$$

Therefore the lower part of the polar, of height $\frac{4}{3}(\gamma/(\gamma+1))(M_1^2 - 1)$, is twice as tall as the upper part, of height $\frac{2}{3}(\gamma/(\gamma+1))(M_1^2 - 1)$. The height of the sonic point is three-quarters that of the maximum deflection point, and the deflection at the sonic point is half the maximum deflection.

The formulae we have given in this section are among the most useful and widely applied in high speed flow. More than in the rest of the book, we have aimed at completeness, and it is hoped that the section may serve as a reference for research workers. Many of the formulae embody scaling laws that transcend the theory of shocks and apply generally. For example, in formulae for pressure and deflection angle we have emphasized the limiting cases $M_1 \to \infty$ and $M_1 \to 1$. The various powers of M_1 and $(M_1^2 - 1)^{\frac{1}{2}}$ we have obtained enter into the hypersonic and transonic similarity parameters, and these will reappear in later chapters, for example in Chapter 9 on aerofoils. Theorems about shock polars have been proved for fluids with rather general equation of state, but they are beyond the scope of this book.

6.5 Bibliographic Notes

Early papers on shocks are Rankine (1870), Mach (1878), Hugoniot (1889), Rayleigh (1910), Taylor (1910), Bethe (1942), von Neumann (1943), and others in Table 13.2.1. See also the collection of classic papers on shocks, translated into English, where necessary, in Johnson & Chéret (1998). Reference works on shocks in Table 13.3.1 are Illingworth (1953), Kynch (1953), Ferri (1955b), Hayes (1958), Polachek & Seeger (1958), Cabannes (1960), and Bleakney & Emrich (1961). Most of the texts and monographs in Table 13.4.1 contain chapters on shocks, and particularly thorough treatments are given in Courant & Friedrichs (1948), Zel'dovich & Raizer (1966, 1967), Thompson (1985), Landau & Lifshitz (1987), Ben-Dor (1992), and Glass & Sislian (1994). Of the surveys and reviews listed in Table 13.5.1, those on shocks are Guderley (1953), Lighthill (1956), Zel'dovich & Raizer (1969), Hayes (1971), Adamson & Messiter (1980), Griffith (1981), Lesser & Field (1983), Hornung (1986), Moretti (1987), Ben-Dor (1988), and Menikoff & Plohr (1989). Many excellent photographs of shocks appear in Van Dyke (1982).

Papers on shocks in the *Journal of Fluid Mechanics* in the period 1990–1998 are listed in Table 13.6.1a. The vigorous state of the subject testifies to the importance of shocks in high speed flow and is due in large measure to developments in experimental technique and to the increasing powers of computational fluid dynamics, not least in "shock-capture." A subject in its own right, reflected in the entries in Table 13.6.1a, is the rarefaction shock, or "negative shock," which we noted can occur, consistently with increase in entropy, in fluids or mixtures with a special equation of state. For theorems about the shock polar of a fluid with a general equation of state, see Henderson & Menikoff (1998). The journal *Shock Waves,* founded in 1991, publishes work on shocks in gases, liquids, solids, and two-phase media and contains many striking photographs. Regular symposia, with published proceedings, are the International Symposium on Shock Waves (formerly the International Symposium on Shock Tubes and Waves), the International Mach Reflection Symposium, and the International Symposium on Military Application of Blast Simulation.

6.6 Further Results and Exercises

In what follows, the ratio of specific heats of a polytropic gas is γ, the variables upstream of a shock have subscript 1 and downstream have subscript 2, the shock angle is ϕ, and the flow deflection is θ.

1. For a normal shock in a polytropic gas, prove the relations

$$\frac{p_2}{p_1} = 1 + \frac{2\gamma}{\gamma + 1}(M_1^2 - 1),$$

$$M_2^2 = \frac{1 + \frac{1}{2}(\gamma - 1)M_1^2}{\gamma M_1^2 - \frac{1}{2}(\gamma - 1)}.$$

(a) Deduce that the maximum pressure on a blunt body moving with Mach
number $M > 1$ into fluid which far ahead is at rest and at pressure p_1
is

$$p_1 \left\{ \frac{(\gamma + 1)^{\gamma+1} \left(\frac{1}{2} M^2\right)^\gamma}{2\gamma M^2 - (\gamma - 1)} \right\}^{\frac{1}{\gamma-1}}.$$

(b) A plane shock is travelling at speed U into a stationary polytropic gas
of sound speed c. Let $M = U/c$. Show that the flow speed behind the
shock is $(2/(\gamma + 1))((M^2 - 1)/M)c$.

2. A semi-infinite mass of a polytropic gas has initially a uniform density and
a uniform sound speed c and is at rest. The rigid plane that bounds the gas
is suddenly made to move into the gas at a maintained speed U, causing a
shock to propagate ahead of the plane at a speed V. The conditions behind
the shock are uniform. Proving any normal shock relations you use, find an
expression for V. Show that the passage of the shock increases the absolute
temperature of the gas by the ratio

$$1 + \frac{(\gamma - 1)^2}{4} \frac{U^2}{c^2} + (\gamma - 1)\frac{U}{c}\left\{ 1 + \left\{ \left(\frac{\gamma+1}{4}\right)\frac{U}{c}\right\}^2 \right\}^{\frac{1}{2}}.$$

3. A normal shock propagates into a stationary polytropic gas in which the
speed of sound is c. The ratio of the pressure behind the shock to the
pressure in the stationary gas ahead of the shock is λ. Given that $\lambda > 1$,
prove that the propagation speed of the shock into the stationary gas is

$$\left\{ 1 + \left(\frac{\gamma + 1}{2\gamma}\right)(\lambda - 1)\right\}^{\frac{1}{2}} c.$$

Explain the significance of the condition $\lambda > 1$.

4. A plane shock in a polytropic gas is moving in a direction normal to itself
and towards a plane rigid wall parallel to the shock. Show that, when
the strength of the shock is large, the excess pressure on the wall after
the shock has been reflected is approximately proportional to the pressure
difference across the incident shock and that the constant of proportionality
is $(3\gamma - 1)/(\gamma - 1)$.

5. A polytropic gas flows through a stationary normal shock. On the two sides
of the shock the pressures are p_1 and p_2, the densities are ρ_1 and ρ_2, and
the velocities are u_1 and u_2. Write down the relations between p_1 and p_2,
between ρ_1 and ρ_2, and between u_1 and u_2. Define the critical speed $\hat{c}$ of

the gas, and obtain Prandtl's relation in the form

$$u_1 u_2 = \frac{p_2 - p_1}{\rho_2 - \rho_1} = \hat{c}^2.$$

Determine the maximum value of ρ_2/ρ_1 that can be achieved by the shock.

6. A strong normal shock is advancing at speed U into a stationary polytropic gas that has pressure p_1, density ρ_1, and speed of sound c_1. Behind the shock, the gas has pressure p_2, density ρ_2, and speed of sound c_2. Show that, approximately,

$$\frac{\rho_2}{\rho_1} = \frac{\gamma + 1}{\gamma - 1}, \qquad \frac{p_2}{p_1} = \frac{2\gamma}{\gamma + 1}\frac{U^2}{c_1^2}, \qquad \frac{u_2}{U} = \frac{2}{\gamma + 1}.$$

A strong cylindrical shock in a polytropic gas is created by the explosion of a line charge that releases an amount of energy per unit length E. Show that, while the shock remains strong, the radius R of the shock after time t from the initial instant is given in terms of the density ρ_0 of the undisturbed gas and a constant A, which need not be evaluated, by

$$R = A\left(\frac{E}{\rho_0}\right)^{\frac{1}{4}} t^{\frac{1}{2}}.$$

Explain briefly how this result can be applied to the determination of the shape of the shock behind an axially symmetric body moving uniformly at hypersonic speed.

7. State the conditions that must be satisfied by a polytropic gas on the two sides of a straight stationary shock.

If the ratio of the pressures on the two sides of a stationary normal shock is $1 + \epsilon$, where ϵ is small, what are the ratios of the densities and the velocities, to order ϵ? Determine the change in entropy to the same order in ϵ, and comment on your result.

By considering higher powers of ϵ, deduce that the Mach number ahead of the shock is

$$1 + \left(\frac{\gamma + 1}{4\gamma}\right)\epsilon + O(\epsilon^2).$$

8. Obtain the equations relating the conditions on the two sides of a steady shock in a polytropic gas. Show that although the ratio of the temperatures on the two sides can be arbitrarily large, the ratio of the densities is bounded.

Let the upstream Mach number be $1 + \epsilon$, where $|\epsilon| \ll 1$. Find the change in entropy correct to order ϵ^3, and deduce the direction of variation of temperature and density across the shock.

9. Obtain the Rankine–Hugoniot relation between the pressure, density, and velocity on opposite sides of a normal shock. Show that, if the shock is at rest, the flow must be subsonic on one side of the shock and supersonic on the other.

10. A stationary plane oblique shock is inclined at an angle ϕ to an oncoming stream of a polytropic gas having speed of sound c, flow speed q, and Mach number M. After passing through the shock, the gas has velocity components u parallel to the original direction of flow and v perpendicular to it. Prove that

$$v^2 = \frac{(q-u)^2\{(\gamma+1)qu - (\gamma-1)q^2 - 2c^2\}}{2c^2 + (\gamma+1)q(q-u)}.$$

You may quote without proof any formulae relating to normal shocks, but you should give in full all subsequent calculations.

Deduce that if the deflection of the stream is small, and the shock is weak, then

$$\tan\phi \simeq (M^2 - 1)^{-\frac{1}{2}}.$$

Give a simple explanation of this result.

11. Obtain the Rankine–Hugoniot relation for a stationary oblique shock in a polytropic gas.

A uniform stream of a polytropic gas with speed u_1, pressure p_1, density ρ_1, and speed of sound c_1 contains an oblique shock at an angle ϕ. Upstream, the component of flow velocity normal to the shock is u_{1n}, and downstream it is u_{2n}. Show that

$$u_{1n}u_{2n} = \left(\frac{\gamma-1}{\gamma+1}\right)u_1^2 \sin^2\phi + \left(\frac{2}{\gamma+1}\right)c_1^2.$$

The upstream Mach number is $M_1 = u_1/c_1$, and the deflection of the stream on passing through the shock is θ. Show that

$$\tan\theta = \frac{(M_1^2 \sin^2\phi - 1)\cot\phi}{1 + \frac{1}{2}M_1^2(\gamma + \cos 2\phi)}.$$

A stream for which $M_1\theta \gg 1$ is called hypersonic. Show that the shock angle of a hypersonic stream satisfies approximately the relation $\phi = \frac{1}{2}(\gamma+1)\theta$ and that the normal velocity, pressure, and density behind the shock are approximately

$$\left(\frac{\gamma-1}{2}\right)u_1\theta, \qquad p_1 + \left(\frac{\gamma+1}{2}\right)\rho_1 u_1^2\theta^2, \qquad \left(\frac{\gamma+1}{\gamma-1}\right)\rho_1.$$

12. State the Rankine–Hugoniot relation connecting the flow conditions on opposite sides of a stationary normal shock in a polytropic gas. Show that, if the gas has upstream density ρ_1, downstream density ρ_2, and upstream Mach number M_1, then

$$\frac{\rho_2}{\rho_1} = \frac{\frac{1}{2}(\gamma + 1)M_1^2}{1 + \frac{1}{2}(\gamma - 1)M_1^2}.$$

Now suppose that the shock is inclined to the upstream direction at an angle ϕ and that $M_1 \gg 1$. Show that the flow deflection θ produced by the shock satisfies approximately the relation

$$\tan \theta = \frac{\sin 2\phi}{\gamma + \cos 2\phi}.$$

Find the maximum value of θ.

13. Prove that at high Mach numbers the maximum flow deflection possible at a plane stationary oblique shock is $\tan^{-1}((\gamma^2 - 1)^{-\frac{1}{2}})$ and that the angle between the shock and the incident stream is then $\cos^{-1}(((\gamma - 1)/(2\gamma))^{\frac{1}{2}})$.

14. A stationary plane shock lies obliquely in an incident stream of a polytropic gas of speed u_1 and sound speed c_1. Behind the shock, the velocity component parallel to the incident stream is u and the component perpendicular to it is v. Show that

$$\left\{ u - u_1 - \frac{2c_1^2}{(\gamma + 1)u_1} \right\} \{(u - u_1)^2 + v^2\} + \frac{2u_1}{\gamma + 1}(u - u_1)^2 = 0.$$

Determine the pressure difference across the shock, and deduce that the largest angle through which a shock can turn a stream is $\sin^{-1}(1/\gamma)$.

7

Steady One-Dimensional Flow

7.1 Flow in a Stream Tube

In this section we begin with a stream tube of an arbitrary ideal fluid, and then we consider in more detail a stream tube of a polytropic gas. In the rest of the chapter we apply the results to flows in ducts and nozzles.

The cross-sectional area A of the stream tube is assumed to vary sufficiently slowly with distance that the flow depends only on distance along the tube and so is approximately one dimensional, as in Figure 7.1.1. We ignore body forces and take the flow to be steady, isentropic, and free of shocks. The fluid has speed u, density ρ, pressure p, temperature T, enthalpy h, speed of sound c, and Mach number M, all of which vary slowly along the stream tube. Conservation of mass and Bernoulli's equation give

$$\rho u A = \text{const.}, \tag{7.1.1a}$$

$$h + \tfrac{1}{2}u^2 = \text{const.} \tag{7.1.1b}$$

In differentials, these equations are equivalent to

$$\frac{d\rho}{\rho} + \frac{du}{u} + \frac{dA}{A} = 0, \tag{7.1.2a}$$

$$dh + u\, du = 0. \tag{7.1.2b}$$

Since the flow is isentropic, we have $dp = c^2 d\rho$ and $dh = \rho^{-1}dp = \rho^{-1}c^2 d\rho$, so that (7.1.2b) may be written in the equivalent forms

$$dp + \rho u\, du = 0, \tag{7.1.3a}$$

$$\frac{d\rho}{\rho} + \frac{u\, du}{c^2} = 0. \tag{7.1.3b}$$

Eliminating $\rho^{-1}d\rho$ between (7.1.2a) and (7.1.3b) gives $A^{-1}dA = c^{-2}u\, du -$

120

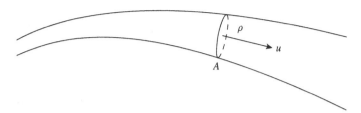

Fig. 7.1.1. Flow in stream tube. Cross-sectional area A, fluid density ρ, and flow speed u, depending only on distance along the tube.

$u^{-1}du$, which can be expressed in terms of the Mach number $M = u/c$ as

$$\frac{dA}{A} = (M^2 - 1)\frac{du}{u}. \tag{7.1.4}$$

Therefore while the flow is subsonic, an increase in u occurs simultaneously with a decrease in A (i.e., $du > 0$ implies $dA < 0$), but once the flow has become supersonic, an increase in u occurs with an increase in A (i.e., $du > 0$ implies $dA > 0$). Thus as a compressible fluid accelerates, the cross-sectional area of a stream tube decreases to a minimum value where the flow is sonic, that is, where the flow speed equals the local sound speed, and thereafter the cross-sectional area increases. This fact is of great significance in high speed flow, because it implies that fluid cannot be accelerated to supersonic speed unless it is given "room to expand sideways." Hence in a duct of uniform cross section that opens at one end into stationary fluid, no amount of increase in pressure at the other end can induce the fluid to issue from the duct supersonically. The physical explanation is that reduction in pressure has two separate effects. The first of these is to accelerate the fluid, an effect that tends to reduce the cross-sectional area, because of conservation of mass; the second is to allow the fluid to expand, an effect that tends to increase the cross-sectional area. At subsonic speeds, the former effect wins, whereas at supersonic speeds, the latter dominates. The factor $M^2 - 1$ in (7.1.4) represents the competition between the two effects.

An alternative form of (7.1.3a) is

$$\frac{dp}{\rho c^2} = -M^2 \frac{du}{u}, \tag{7.1.5}$$

and an alternative form of (7.1.4), obtained by writing the equation $\rho u A = $ constant as $(\rho u)^{-1}d(\rho u) + A^{-1}dA = 0$, is

$$\frac{d(\rho u)}{\rho u} = -(M^2 - 1)\frac{du}{u}. \tag{7.1.6}$$

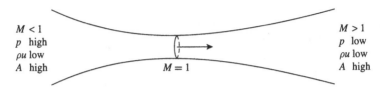

$M < 1$
p high
ρu low
A high

$M = 1$

$M > 1$
p low
ρu low
A high

Fig. 7.1.2. Acceleration of fluid through sonic speed in a stream tube, and the associated variation in M, p, ρu, and A. The cross-sectional area is a minimum where $M = 1$.

The quantity ρu is the mass of fluid per unit area flowing along the stream tube per unit time (i.e., the mass flux). Therefore, during acceleration, the mass flux increases to a maximum value where the flow is sonic, and thereafter it decreases. The changes in M, p, ρu, and A as fluid accelerates through a sonic throat are summarised in Figure 7.1.2. The most important consequence of (7.1.4) is that if $M = 1$ then $dA = 0$ (or if $dA \neq 0$ then $M \neq 1$). The only places where the fluid can attain sonic speed are where the cross-sectional area of a stream tube is stationary.

Now assume that the fluid is a polytropic gas with gas constant R and ratio of specific heats γ. Then $p = R\rho T$ and $h = c^2/(\gamma - 1) = \gamma RT/(\gamma - 1)$. Constant entropy gives $p \propto \rho^\gamma$. Therefore in the stream tube we have

$$\rho u A = \text{const.,} \tag{7.1.7a}$$

$$\frac{c^2}{\gamma - 1} + \frac{u^2}{2} = \text{const.,} \tag{7.1.7b}$$

$$p\rho^{-\gamma} = \text{const.} \tag{7.1.7c}$$

Since Bernoulli's equation (7.1.7b) may be written $\gamma RT/(\gamma - 1) + \frac{1}{2}u^2 = $ constant, it expresses the fact that as the speed goes up the temperature goes down, and vice versa. This describes precisely the interchange between kinetic and thermal energy as fluid accelerates or decelerates. The interchange is due entirely to the fluid's compressibility and has no counterpart in incompressible flow.

The Bernoulli constant on the right of (7.1.7b) may be written in terms of certain parameters of the flow in three equivalent ways by evaluating the left-hand side at three reference conditions: the stagnation condition, $u = 0, c = c_0$; the critical condition, $u = c = \hat{c}$; and the vacuum condition, $u = u_{\text{max}}, c = 0$. That is,

$$\frac{c^2}{\gamma - 1} + \frac{u^2}{2} = \frac{c_0^2}{\gamma - 1} = \left(\frac{\gamma + 1}{\gamma - 1}\right)\frac{\hat{c}^2}{2} = \frac{u_{\text{max}}^2}{2}. \tag{7.1.8}$$

Thus c_0 is the stagnation sound speed; $\hat{c}$ is the critical speed, at which the flow is sonic; and u_{max} is the maximum speed, attained when the absolute temperature

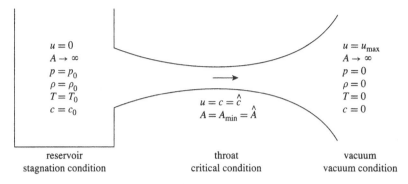

Fig. 7.1.3. Flow from a reservoir, through a throat, and into a vacuum.

is zero. In this latter case, the whole of the disordered thermal energy of the fluid has been converted to ordered kinetic energy, and we may imagine every molecule of the fluid travelling in a straight line at the speed u_{max}. Recall that at constant entropy we have $\rho \propto c^{2/(\gamma-1)}$, so that $c \to 0$ corresponds to $\rho \to 0$, that is, a vacuum. The stagnation condition occurs in a plenum, or reservoir; the critical condition can occur only at a position of minimum cross-sectional area A, that is, at a throat; and the vacuum condition is approached at a downstream position of vanishing pressure. We shall see that A increases without limit as the vacuum condition is approached, so that our one-dimensional approximation then becomes inaccurate. A stream tube leaving a reservoir and narrowing to a throat before expanding into a vacuum is shown in Figure 7.1.3.

Equation (7.1.8) gives

$$u_{max} = \left(\frac{2}{\gamma - 1} \right)^{\frac{1}{2}} c_0 = \left(\frac{\gamma + 1}{\gamma - 1} \right)^{\frac{1}{2}} \hat{c}. \tag{7.1.9}$$

For a diatomic gas, with $\gamma = 7/5$, this is

$$u_{max} = \sqrt{5}\, c_0 = \sqrt{6}\, \hat{c}. \tag{7.1.10}$$

For a polytropic gas at constant entropy, the four quantities p, ρ^γ, $T^{\gamma/(\gamma-1)}$, and $c^{2\gamma/(\gamma-1)}$ vary in proportion. Since $\hat{c}/c_0 = (2/(\gamma+1))^{\frac{1}{2}}$, it follows that critical conditions $\hat{p}$, $\hat{\rho}$, $\hat{T}$, $\hat{c}$ are related to stagnation conditions p_0, ρ_0, T_0, c_0 by

$$\left(\frac{\hat{p}}{p_0}, \frac{\hat{\rho}}{\rho_0}, \frac{\hat{T}}{T_0}, \frac{\hat{c}}{c_0} \right) = \left(\left(\frac{2}{\gamma+1} \right)^{\frac{\gamma}{\gamma-1}}, \left(\frac{2}{\gamma+1} \right)^{\frac{1}{\gamma-1}}, \frac{2}{\gamma+1}, \left(\frac{2}{\gamma+1} \right)^{\frac{1}{2}} \right).$$

$$\tag{7.1.11}$$

For a diatomic gas, with $\gamma = 7/5$, this is

$$\left(\frac{\hat{p}}{p_0}, \frac{\hat{\rho}}{\rho_0}, \frac{\hat{T}}{T_0}, \frac{\hat{c}}{c_0}\right) = \left(\left(\frac{5}{6}\right)^{\frac{7}{2}}, \left(\frac{5}{6}\right)^{\frac{5}{2}}, \frac{5}{6}, \left(\frac{5}{6}\right)^{\frac{1}{2}}\right)$$

$$= (0.528, 0.634, 0.833, 0.913).$$

(7.1.12)

We may write (7.1.7), and the equation of state $p = R\rho T$, in terms of the Mach number $M = u/c$. The result, reflecting the proportionality of p, ρ^γ, $T^{\gamma/(\gamma-1)}$, and $c^{2\gamma/(\gamma-1)}$, is

$$\frac{M}{\left(1 + \frac{1}{2}(\gamma - 1)M^2\right)^{\frac{1}{2}\left(\frac{\gamma+1}{\gamma-1}\right)}} A = \text{const.}, \qquad \frac{\left(1 + \frac{1}{2}(\gamma - 1)M^2\right)^{\frac{1}{2}}}{M} u = \text{const.},$$

$$\left(1 + \frac{1}{2}(\gamma-1)M^2\right)^{\frac{\gamma}{\gamma-1}} p = \text{const.}, \quad \left(1 + \frac{1}{2}(\gamma-1)M^2\right)^{\frac{1}{\gamma-1}} \rho = \text{const.}, \quad (7.1.13)$$

$$\left(1 + \frac{1}{2}(\gamma - 1)M^2\right) T = \text{const.}, \quad \left(1 + \frac{1}{2}(\gamma - 1)M^2\right)^{\frac{1}{2}} c = \text{const.}$$

For a diatomic gas, with $\gamma = 7/5$, the four proportional quantities are p, $\rho^{7/5}$, $T^{7/2}$, and c^7, and Equations (7.1.13) are

$$\frac{M}{\left(1 + \frac{1}{5}M^2\right)^3} A = \text{const.}, \qquad \frac{\left(1 + \frac{1}{5}M^2\right)^{\frac{1}{2}}}{M} u = \text{const.},$$

$$\left(1 + \frac{1}{5}M^2\right)^{\frac{7}{2}} p = \text{const.}, \quad \left(1 + \frac{1}{5}M^2\right)^{\frac{5}{2}} \rho = \text{const.}, \qquad (7.1.14)$$

$$\left(1 + \frac{1}{5}M^2\right) T = \text{const.}, \quad \left(1 + \frac{1}{5}M^2\right)^{\frac{1}{2}} c = \text{const.},$$

A check shows that when A satisfies (7.1.13), that is, when A is proportional to $M^{-1}(1 + \frac{1}{2}(\gamma-1)M^2)^{(\gamma+1)/(2(\gamma-1))}$, its minimum value $A_{\min}$ is attained when $M = 1$, in accordance with the theory after (7.1.4), so that $A_{\min}$ may equally be called the critical cross-sectional area and written $\hat{A}$. Using $M = 1$ and $A = A_{\min}$ or $A = \hat{A}$ to evaluate the constant in (7.1.13) gives

$$\frac{A}{A_{\min}} = \frac{A}{\hat{A}} = \left(\frac{2}{\gamma + 1}\right)^{\frac{1}{2}\left(\frac{\gamma+1}{\gamma-1}\right)} \frac{\left(1 + \frac{1}{2}(\gamma - 1)M^2\right)^{\frac{1}{2}\left(\frac{\gamma+1}{\gamma-1}\right)}}{M}. \qquad (7.1.15)$$

This function of M is slowly varying near $M = 1$, where its graph is thus rather flat, as shown for a diatomic gas with $\gamma = 7/5$ and $A/\hat{A} = (\frac{5}{6})^3(1 + \frac{1}{5}M^2)^3/M$ in Figure 7.1.4. Therefore as fluid accelerates through sonic speed, its Mach number can change substantially with only a small change in the cross-sectional area of the stream tube.

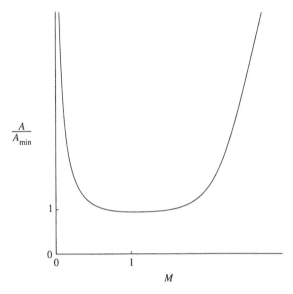

Fig. 7.1.4. Normalised cross-sectional area of a stream tube, $A/A_{\min}$, as a function of Mach number M, for a diatomic gas with $\gamma = \frac{7}{5}$.

7.2 Flow in Ducts and Nozzles. Choked Flow

Consider a stream of fluid escaping from a reservoir through a converging nozzle, as shown in Figure 7.2.1. The stream converges at the end of the nozzle and for a short distance beyond, but thereafter it diverges, so that it has a throat, called a vena contracta, at which the cross-sectional area is a minimum. The expression $\rho u A$ for the mass flow is most accurate when evaluated at the vena contracta, because the flow throughout the cross section is then parallel to the main stream direction. If the expression is evaluated at the end of the nozzle, it must be multiplied by a coefficient that allows for the nonparallel flow (i.e., the flow convergence) to give the mass flow. The value of the coefficient is known in engineering practice for different shapes of nozzle.

Well inside the reservoir, the fluid is at rest, at the stagnation condition p_0, ρ_0, T_0, c_0. At the vena contracta, of cross-sectional area A_1, we shall indicate conditions by a subscript 1 (i.e., u_1, p_1, ρ_1, T_1, c_1). The pressure downstream approaches the ambient pressure p_2. If $p_0 - p_2$ is increased, the Mach number of the flow at the vena contracta increases until the flow is sonic, and at higher $p_0 - p_2$ the flow remains sonic, by the theory in Section 7.1. The flow is then said to be choked. It is therefore impossible to obtain supersonic flow out of a converging nozzle. To produce supersonic flow, a nozzle requires a diverging section, as we shall see in Section 7.3.

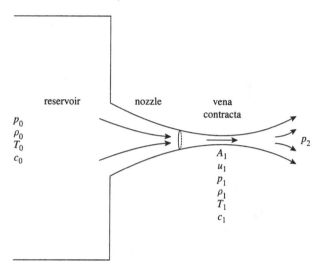

Fig. 7.2.1. Flow from a reservoir at pressure p_0, through a converging nozzle and a vena contracta, to an ambient region at pressure p_2.

The mass of fluid passing through the nozzle per unit time is called the discharge from the reservoir and has the value $\rho_1 u_1 A_1$. The maximum discharge is produced when the flow is choked, that is, when the fluid at the vena contracta is at the critical condition $\hat{u}, \hat{p}, \hat{\rho}, \hat{T}, \hat{c}$, for which $\hat{u} = \hat{c}$ by definition, and so the maximum discharge is $\hat{\rho} \hat{c} A_1$. Assume that the reservoir contains a diatomic gas, with ratio of specific heats $\gamma = 7/5$, at known conditions. Then by (7.1.12), the choked conditions at the end of the nozzle are

$$(u, p, \rho, T, c, \rho u A) = (\hat{u}, \hat{p}, \hat{\rho}, \hat{T}, \hat{c}, \hat{\rho}\hat{c}A_1)$$

$$= \left(\left(\frac{5}{6}\right)^{\frac{1}{2}} c_0, \left(\frac{5}{6}\right)^{\frac{7}{2}} p_0, \left(\frac{5}{6}\right)^{\frac{5}{2}} \rho_0, \left(\frac{5}{6}\right) T_0, \left(\frac{5}{6}\right)^{\frac{1}{2}} c_0, \left(\frac{5}{6}\right)^{3} \rho_0 c_0 A_1 \right)$$

$$= (0.913 c_0, 0.528 p_0, 0.634 \rho_0, 0.833 T_0, 0.913 c_0, 0.579 \rho_0 c_0 A_1). \quad (7.2.1)$$

Thus the flow becomes choked when the ambient pressure is about half the reservoir pressure. A simple explanation of the fact that ambient conditions have no further effect once the flow at the end of the nozzle has become sonic is that disturbances cannot travel backwards against a sonic or supersonic stream.

The variation of u, p, ρ, T, c, and M along a nozzle with a given variation of A along its length is determined in terms of A by Equations (7.1.13) or, if $\gamma = 7/5$, by (7.1.14). A nozzle is a type of duct, and these equations may be used generally to describe the one-dimensional flow of a polytropic gas along a duct of varying cross section.

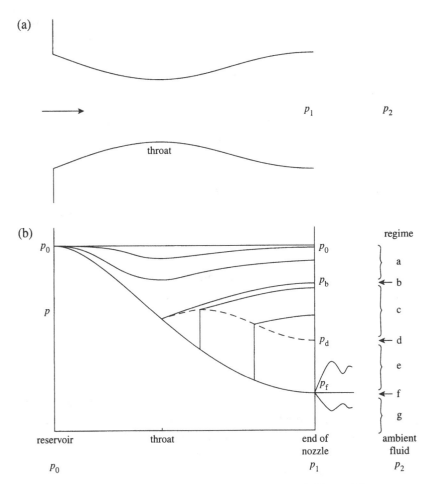

Fig. 7.3.1. The Laval nozzle. (a) Shape of the nozzle, and definitions of the pressures p_1, p_2. (b) Pressure distribution along the nozzle and just beyond its end. Regimes a–g, and transition pressures p_b, p_d, p_f.

7.3 The Laval Nozzle

A nozzle with a diverging section to permit supersonic flow is important in experimental work and is called a *Laval nozzle* or a converging–diverging nozzle. The pressure distribution along such a nozzle depends on the reservoir pressure p_0 and ambient pressure p_2 as shown in Figure 7.3.1. We shall distinguish seven flow regimes, indicated by the letters a, b, c, d, e, f, and g in Figure 7.3.1b. The position of each letter indicates the condition for the occurrence of the regime, that is, the value or range of values of p_2 in relation to p_0.

   The pressure p_1 at the end of the nozzle can differ markedly from p_2, especially in regimes e and g, for which the pressure variation just beyond the end of the nozzle takes the oscillatory form shown by the two curves at the bottom right of Figure 7.3.1b. Regimes b, d, and f correspond to definite values of p_1 or p_2, denoted p_b, p_d, and p_f. These are the downstream pressures that mark the transitions between regimes a, c, e, and g. If p_1 lies in the range $p_f < p_1 < p_b$, then the isentropic stream-tube equations we have been using have no solution, and a shock forms either inside or beyond the diverging part of the nozzle.

We now consider in turn the regimes a–g. They could be produced in this order by steadily reducing p_2 from an initial value equal to p_0.

(a) The Regime $p_b < p_2 < p_0$

Then $p_1 = p_2$, and the flow is subsonic everywhere. In the diverging section there is a smooth decrease in speed and a smooth increase in pressure.

(b) The Regime $p_2 = p_b$

Then $p_1 = p_b = p_2$, and the flow is sonic at the throat of the nozzle. As in regime a, in the diverging section there is a smooth decrease in speed and a smooth increase in pressure.

(c) The Regime $p_d < p_2 < p_b$

Then $p_1 = p_2$, and a shock occurs in the diverging section. In Figure 7.3.1b, points on the smooth solid line ending at p_f correspond to the pressure just before the shock, and points on the smooth dashed line ending at p_d correspond to the pressure just after the shock. Thus the vertical lines in the figure give the position and strength of the shock in the diverging section. Therefore when p_2 is only slightly below p_b, the shock is near the throat and not strong. As p_2 is reduced, the shock moves down the diverging section and its strength increases. As p_2 approaches p_d, the shock approaches the end of the nozzle.

(d) The Regime $p_2 = p_d$

Just outside the nozzle, $p_1 = p_d = p_2$; and just inside, $p_1 = p_f$. The shock has reached the end of the nozzle.

(e) The Regime $p_f < p_2 < p_d$

Then $p_1 = p_f < p_2$, and a shock forms outside the nozzle, as shown in Figure 7.3.2 a,b. When p_2 is only slightly below p_d, the shock is close to the end of the nozzle, and most of the shock forms a disc, called a Mach disc.

(a)

(b)

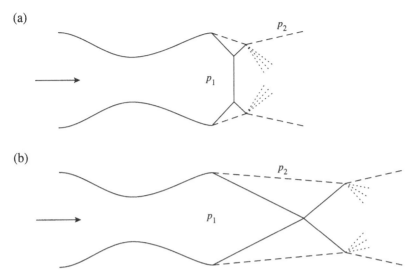

Fig. 7.3.2. Overexpanded jet emerging from a Laval nozzle, corresponding to regime e, for which $p_f < p_2 < p_d$ and $p_1 < p_2$. The figures show shocks (——), the jet boundary (– – –), and Prandtl–Meyer expansion fans (·····). (a) p_2 slightly below p_d; (b) p_2 slightly above p_f. In (a), the vertical line is a diameter of the Mach disc, between Mach triple points.

The rim of the disc is a line of triple points of intersection of shocks, called Mach triple points, where the disc is attached to two oblique shocks. The first of these is attached at its other end to the rim of the nozzle. The second meets the vortex sheet at the boundary between ambient fluid and the jet of fluid issuing from the nozzle and is then reflected as a fan-shaped expansion wave, called a Prandtl–Meyer expansion, or an expansion fan. Figure 7.3.2a shows the configuration of shocks for p_2 slightly below p_d. As p_2 is reduced, the Mach disc moves further away from the end of the nozzle, and its radius decreases, until below a certain value of p_2 the Mach disc is no longer present, and the shocks from the rim of the nozzle simply meet or reflect off each other, at a point of regular intersection or reflection. Reduction of p_2 from p_d to p_f causes a steady reduction in the strength of the shocks. The configuration of shocks for p_2 well below p_d, but still slightly above p_f, is shown in Figure 7.3.2b. In the regime we are describing, the fluid in the diverging section has expanded so much that the pressure p_1 at the end of the nozzle is less than the pressure p_2 of the ambient fluid, and the shocks provide the mechanism for bringing the jet pressure up to the ambient pressure. In this regime, the jet is said to be overexpanded. As can be seen in Figure 7.3.2, the flow in an overexpanded jet is rather complicated. In later chapters we shall say more about the flow features we have introduced, namely the Mach disc and triple

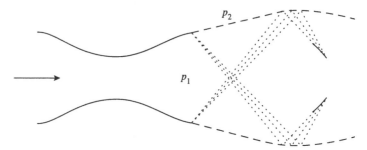

Fig. 7.3.3. Underexpanded jet emerging from a Laval nozzle, corresponding to regime g, for which $p_2 < p_f$ and $p_1 > p_2$. The figure shows Prandtl–Meyer expansion fans and their reflections ($\cdots\cdots$), the jet boundary ($-\,-\,-$), and shocks ($-\!-\!-$).

point, the regular intersection and reflection of shocks, and the Prandtl–Meyer expansion fan.

(f) The Regime $p_2 = p_f$

Then $p_1 = p_f = p_2$, and the shocks found in regime e now have zero strength (i.e., they have become Mach waves). In this regime, the jet is said to be fully expanded, and the Laval nozzle is operating at its design condition.

(g) The Regime $p_2 < p_f$

Then $p_1 = p_f > p_2$, and outside the nozzle there is a Prandtl–Meyer expansion fan, as shown in Figure 7.3.3. The fan reflects off the vortex sheet that forms the boundary of the jet, and, as shown on the right of Figure 7.3.3, the reflection of the fan can converge to form a shock. In this regime, the fluid in the diverging section has not expanded enough to bring the pressure p_1 of the fluid at the end of the nozzle down to the pressure p_2 of the ambient fluid. The Prandtl–Meyer expansion is the mechanism for bringing the pressure down. In the regime $p_2 < p_f$, the jet is said to be underexpanded.

The above description of the various regimes of flow in a Laval nozzle, though complicated, is still somewhat simplified. We have not mentioned (i) localised supersonic regions in the throat; (ii) flow separation in the diverging section; (iii) boundary layer/shock interactions, which can produce multiple shocks in the diverging section; (iv) successive reflections of shocks and expansions off the boundary of the jet, and the resulting cellular structure of shocks in the jet; and (v) complicated unsteady motions, including shock oscillations, often leading to intense sound production, for example "screech." These complications are important in practice and have been intensively studied, not least because a large nozzle is a wind tunnel, and we have introduced in this chapter some ideas that are fundamental in wind tunnel engineering.

7.4 Bibliographic Notes

Reference works in Table 13.3.1 that cover steady one-dimensional flow, including flow in nozzles and wind tunnels, are Saunders (1953), Evvard (1957), Crocco (1958), Goddard (1961), and Smith (1961). These works, and also the texts and monographs in Table 13.4.1, include the effect of heat transfer between the fluid and the wall of the duct, as described by the theory that leads to the Fanno line and Rayleigh line. In the research papers in Table 13.6.1d, for the flow of a gas–particle mixture in a duct see Buresti & Casarosa (1993); for transonic nozzle flow see Kluwick (1993); and for low Mach number channel flow see Shajii & Freidberg (1996). Difficult mathematical work on high speed flow in ducts and nozzles was performed in the 1940s and 1950s, of which a representative example is Cherry (1950). Some of this work found its way into texts and monographs, for example Bers (1958).

7.5 Further Results and Exercises

1. A polytropic gas with ratio of specific heats γ is in steady one-dimensional flow at a stagnation temperature T_0. Show that the temperature T of the gas, as a function of the Mach number M of the flow, is given by $T/T_0 = (1 + \frac{1}{2}(\gamma - 1)M^2)^{-1}$. The stream tube area A takes the value $\hat{A}$ when $M = 1$. Show that

$$\left(\frac{A}{\hat{A}}\right)^{2\left(\frac{\gamma-1}{\gamma+1}\right)} = \left(\frac{\gamma-1}{\gamma+1}\right)M^{\frac{4}{\gamma+1}} + \left(\frac{2}{\gamma+1}\right)M^{-2\left(\frac{\gamma-1}{\gamma+1}\right)}.$$

Sketch the graph of the relation between A and M. Verify, by an accurate plot, the flatness of the graph over a wide range of M containing $M = 1$.

2. Write an essay on temperature variations in fluid moving at high speed. Include detailed calculations for simple examples, such as flow in a stream tube or through a shock wave.

3. A polytropic gas with ratio of specific heats γ flows adiabatically from a vessel where the pressure is p_0 through a small aperture to a region where the pressure is p_1. The issuing jet contracts a short distance from the aperture to a minimum cross section of area A_1, where the fluid has density ρ_1 and speed of sound c_1. Show that the mass flow rate from the vessel is

$$\rho_1 c_1 A_1 \left(\frac{2}{\gamma - 1}\right)^{\frac{1}{2}} \left\{ \left(\frac{p_0}{p_1}\right)^{\frac{\gamma-1}{\gamma}} - 1 \right\}^{\frac{1}{2}}.$$

Deduce that if the conditions at the minimum cross section are sonic, then

$$\frac{p_0}{p_1} = \left(\frac{\gamma + 1}{2}\right)^{\frac{\gamma}{\gamma-1}}.$$

What is the value of this expression if the gas is (i) air; (ii) helium?

4. A perfect gas with ratio of specific heats γ flows steadily through a nozzle. The gas has stagnation pressure p_0, and the flow is one dimensional and homentropic. Write down a set of equations that determines the variation of pressure p and Mach number M along the nozzle, and show that

$$\frac{p}{p_0} = \left(1 + \frac{1}{2}(\gamma - 1)M^2\right)^{-\frac{\gamma}{\gamma-1}}.$$

A supersonic wind tunnel, of throat area $\hat{A}$, is of length $3x_0$ between throat and exit and, at a distance x downstream of the throat, has cross-sectional area $A = (1 + (x/x_0)^2)\hat{A}$. The wind tunnel is operating without the production of shocks. Does the flow attain a local Mach number of 2 within the wind tunnel?

5. State the equations that describe the one-dimensional motion of an ideal polytropic gas with ratio of specific heats γ. At position x and time t, the gas has density $\rho(x, t)$, speed $u(x, t)$, and speed of sound $c(x, t)$. The undisturbed gas has density ρ_0 and speed of sound c_0. Show that if the density is a function of speed, then

$$\rho = \rho_0\left(1 + \frac{1}{2}(\gamma - 1)\frac{u}{c_0}\right)^{\frac{2}{\gamma-1}}.$$

Hence show that, for some function $f(u)$, the speed $u(x, t)$ is given implicitly by

$$x = \left(\tfrac{1}{2}(\gamma + 1)u \pm c\right) + f(u).$$

The gas is contained in a semi-infinite cylindrical pipe that occupies the region $x > 0$ and is terminated at $x = 0$ by a piston. For $t \leq 0$, the gas and piston are at rest, and for $t > 0$ the piston has constant positive acceleration α. Determine the function $f(u)$.

6. Write an essay on the differences between subsonic and supersonic flow of a gas in stream tubes, ducts, and nozzles.

8

Prandtl–Meyer Expansion

8.1 Problems in Two Independent Variables

Most flows that lead to problems in two independent variables are of the types (a) unsteady one-dimensional; (b) unsteady spherically symmetric; (c) steady planar; (d) steady conical; or (e) steady axisymmetric. Examples of these flows are (a) piston problems and flow in a shock tube, where typical phenomena of interest are wave steepening and formation of N-waves and shocks; (b) blast waves and explosions, often described by similarity solutions; (c) flow past an aerofoil or cascade, often containing shock interactions and Prandtl–Meyer expansions, and sometimes amenable to the theory of characteristics or a hodograph transformation; (d) the conical flow described by the Taylor–Maccoll similarity solution; and (e) ballistic flow (i.e., flow around missiles). In the rest of this book we concentrate on (c), steady plane flow, and we begin with an analysis of the Prandtl–Meyer expansion.

8.2 Prandtl–Meyer Expansion

In supersonic flow, fluid can expand round a sharp convex corner as shown in Figure 8.2.1, instead of separating at the corner as in incompressible flow. The expansion is confined to a fan-shaped region attached at its vertex to the corner and lies between two regions of supersonic rectilinear flow. This flow pattern, which is rather common and may be seen in many photographs, is called a Prandtl–Meyer expansion, and the fan-shaped region is called an expansion fan, or simply a fan. The boundaries of the fan are radial lines at the upstream and downstream Mach angles to the adjacent upstream and downstream flow. The flow pattern admits a complete and simple mathematical description, which we now present.

In what follows, the fluid has pressure p, density ρ, temperature T, speed of sound c, and velocity $\mathbf{u}$. The fluid is assumed ideal and homentropic, so that

133

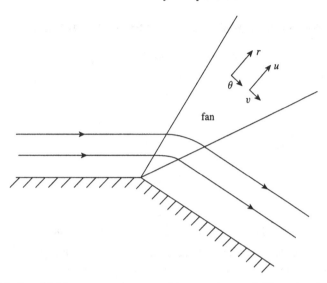

Fig. 8.2.1. Prandtl–Meyer expansion around a corner in a wall. The polar angle θ increases in the clockwise direction.

$dp = c^2 d\rho$ and $\nabla p = c^2 \nabla \rho$. For steady motion, the equations of conservation of mass and momentum are

$$\nabla \cdot (\rho \mathbf{u}) = 0,$$

$$\mathbf{u} \cdot \nabla \mathbf{u} = \frac{-\nabla p}{\rho} = -c^2 \frac{\nabla \rho}{\rho}. \qquad (8.2.1)$$

In the fan region, we use polar coordinates (r, θ) with θ increasing in the clockwise direction, and we denote the corresponding radial and azimuthal components of $\mathbf{u}$ by (u, v), as shown in Figure 8.2.1. In terms of these velocity components, (8.2.1) is

$$\frac{\partial}{\partial r}(\rho r u) + \frac{\partial}{\partial \theta}(\rho v) = 0,$$

$$u\frac{\partial u}{\partial r} + \frac{v}{r}\frac{\partial u}{\partial \theta} - \frac{v^2}{r} = -\frac{1}{\rho}\frac{\partial p}{\partial r} = -\frac{c^2}{\rho}\frac{\partial \rho}{\partial r}, \qquad (8.2.2)$$

$$u\frac{\partial v}{\partial r} + \frac{v}{r}\frac{\partial v}{\partial \theta} + \frac{uv}{r} = -\frac{1}{\rho r}\frac{\partial p}{\partial \theta} = -\frac{c^2}{\rho r}\frac{\partial \rho}{\partial \theta}.$$

We look for a solution of (8.2.2) in which none of the dependent variables depend on r, that is, in which u, v, p, ρ, and c are functions only of θ. Denoting differentiation with respect to θ by $'$, we obtain equations for $u(\theta)$, $v(\theta)$, $p(\theta)$, $\rho(\theta)$, and $c(\theta)$ in the form

$$\rho(u + v') + \rho'v = 0, \tag{8.2.3a}$$

$$v(u' - v) = 0, \tag{8.2.3b}$$

$$v(u + v') = -\frac{p'}{\rho} = -\frac{c^2\rho'}{\rho}. \tag{8.2.3c}$$

Assume that there is some flow and pressure gradient:

$$v \neq 0, \quad p' \neq 0. \tag{8.2.4}$$

Elimination of ρ'/ρ from (8.2.3a,c) gives

$$\frac{u + v'}{v} = \frac{v}{c^2}(u + v'). \tag{8.2.5}$$

Therefore

$$(c^2 - v^2)(u + v') = 0. \tag{8.2.6}$$

Since $p' \neq 0$, it follows from (8.2.3c) that $u + v' \neq 0$. Therefore $c^2 - v^2 = 0$ (i.e., $v = \pm c$). The flow in Figure 8.2.1 is from left to right, so that the convention that θ increases in the clockwise direction gives $v = c$. Equation (8.2.3b) gives $u' = v$. Thus

$$u' = v = c. \tag{8.2.7}$$

A check shows that if we assume (8.2.7), then Equations (8.2.3) have a solution. Equations (8.2.7) are equivalent to the statement that in the fan region the radial lines from the corner O are characteristics for acoustic propagation (i.e., they are Mach lines). This is a consequence of the theory of characteristic surfaces and rays given in Chapter 5, which showed that the relative velocity component perpendicular to a Mach line is the speed of sound, and it may be checked independently by applying the theory to the set of Equations (8.2.2).

The analysis so far applies to an arbitrary fluid. Now assume that the fluid is a polytropic gas with ratio of specific heats γ. In (7.1.8) we gave three expressions for the constant term in Bernoulli's equation for a polytropic gas. We shall use the expression in $u_{\max}$, the speed the fluid would have when $c = 0$. Thus

$$\frac{c^2}{\gamma - 1} + \frac{u^2 + v^2}{2} = \frac{u_{\max}^2}{2}. \tag{8.2.8}$$

With $c = u'$ and $v = u'$, this is an ordinary differential equation for $u(\theta)$:

$$\left(\frac{\gamma + 1}{\gamma - 1}\right) u'^2 + u^2 = u_{max}^2. \tag{8.2.9}$$

The solution contains an arbitrary constant which we incorporate into the definition of θ by taking $u = 0$ when $\theta = 0$. Then

$$u = u_{max} \sin\left\{\left(\frac{\gamma - 1}{\gamma + 1}\right)^{\frac{1}{2}} \theta\right\}. \tag{8.2.10}$$

Therefore

$$v = c = u' = \left(\frac{\gamma - 1}{\gamma + 1}\right)^{\frac{1}{2}} u_{max} \cos\left\{\left(\frac{\gamma - 1}{\gamma + 1}\right)^{\frac{1}{2}} \theta\right\}. \tag{8.2.11}$$

The Mach number M in the fan region is the function of θ determined by $M^2 = (u^2 + v^2)/c^2 = 1 + u^2/v^2$. Thus

$$M^2 = 1 + \left(\frac{\gamma + 1}{\gamma - 1}\right) \tan^2\left\{\left(\frac{\gamma - 1}{\gamma + 1}\right)^{\frac{1}{2}} \theta\right\}, \tag{8.2.12}$$

or

$$(M^2 - 1)^{\frac{1}{2}} = \left(\frac{\gamma + 1}{\gamma - 1}\right)^{\frac{1}{2}} \tan\left\{\left(\frac{\gamma - 1}{\gamma + 1}\right)^{\frac{1}{2}} \theta\right\}. \tag{8.2.13}$$

Therefore the polar angle θ is given in terms of M by

$$\theta = \left(\frac{\gamma + 1}{\gamma - 1}\right)^{\frac{1}{2}} \tan^{-1}\left\{\left(\frac{\gamma - 1}{\gamma + 1}\right)^{\frac{1}{2}} (M^2 - 1)^{\frac{1}{2}}\right\}. \tag{8.2.14}$$

Thus when $\theta = 0$, we have $u = 0$, $v = c = \{(\gamma - 1)/(\gamma + 1)\}^{\frac{1}{2}} u_{max}$, and $M = 1$.

Consider the extension of the fan region in Figure 8.2.1 to the left as far as the radial line $\theta = 0$, which we shall call the reference line. A hypothetical stream of fluid, which to the left of the reference line flows perpendicular to it at Mach number $M = 1$, and which to the right has the flow pattern shown in Figure 8.2.2, containing an extended fan, is called the reference stream. In the right-angled triangle containing (u, v), the angle opposite u is the flow-components

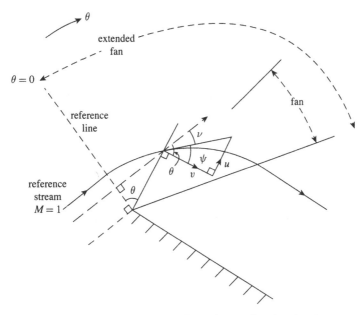

Fig. 8.2.2. Definitions of the fan, extended fan, reference line $\theta = 0$, reference stream, flow components u, v, flow-components angle ψ, and reference-stream deflection angle ν. The Mach angle is $\frac{1}{2}\pi - \psi$. Upstream of the extended fan, the reference stream is sonic and rectilinear by definition. All parts of the figure to the left of the fan are mathematical constructions; the real flow, shown in Figure 8.2.1, is bounded by a wall that is horizontal to the left of the corner.

angle ψ. Thus $\tan \psi = u/v$, so that

$$
\tan \psi = \left(\frac{\gamma + 1}{\gamma - 1} \right)^{\frac{1}{2}} \tan \left\{ \left(\frac{\gamma - 1}{\gamma + 1} \right)^{\frac{1}{2}} \theta \right\}
$$

$$
= (M^2 - 1)^{\frac{1}{2}}.
$$

(8.2.15)

Hence $\psi = \cos^{-1}(1/M)$, that is, ψ is the complement of the Mach angle $\sin^{-1}(1/M)$ corresponding to Mach number M. This also follows directly from the relation $v = c$, which in the velocity-components triangle gives $u^2 + c^2 = M^2 c^2$ (i.e., $u/c = (M^2 - 1)^{\frac{1}{2}}$). Thus in the fan, the flow direction is always at the Mach angle to the radial direction, that is, the radial lines are Mach lines.

The angle turned through by the reference stream is the reference-stream deflection angle ν. Figure 8.2.2 shows that

$$
\nu = \theta - \psi.
$$

(8.2.16)

Therefore

$$v = \theta - \tan^{-1}\left\{\left(\frac{\gamma+1}{\gamma-1}\right)^{\frac{1}{2}} \tan\left\{\left(\frac{\gamma-1}{\gamma+1}\right)^{\frac{1}{2}}\theta\right\}\right\}. \qquad (8.2.17)$$

Using (8.2.14) to write θ and hence v in terms of M, we obtain

$$v(M) = \left(\frac{\gamma+1}{\gamma-1}\right)^{\frac{1}{2}} \tan^{-1}\left\{\left(\frac{\gamma-1}{\gamma+1}\right)^{\frac{1}{2}}(M^2-1)^{\frac{1}{2}}\right\} - \tan^{-1}\left\{(M^2-1)^{\frac{1}{2}}\right\}.$$
$$(8.2.18)$$

The function $v(M)$ is the Prandtl–Meyer function. As M increases from 1 to infinity, the value of $v(M)$ increases from $v(1) = 0$ up to a limiting value v_∞ given by

$$v_\infty = \left\{\left(\frac{\gamma+1}{\gamma-1}\right)^{\frac{1}{2}} - 1\right\}\frac{\pi}{2}. \qquad (8.2.19)$$

In a fan, the deflection of the stream between points where $M = M_1$ and points where $M = M_2$ is $v(M_2) - v(M_1)$. Thus for a stream initially at Mach number M_1, the maximum possible deflection in a Prandtl–Meyer expansion is $v_\infty - v(M_1)$. This expression is greatest when $M_1 = 1$, taking the value v_∞.

The expression for c given in (8.2.11) determines the values of the thermodynamic variables throughout the flow, because of the proportionality, in homentropic flow of a polytropic gas, of the quantities p, ρ^γ, $T^{\gamma/(\gamma-1)}$, and $c^{2\gamma/(\gamma-1)}$. Thus with $\lambda = \{(\gamma - 1)/(\gamma + 1)\}^{1/2}$ we obtain

$$p \propto (\cos \lambda\theta)^{\frac{2\gamma}{\gamma-1}},$$
$$\rho \propto (\cos \lambda\theta)^{\frac{2}{\gamma-1}},$$
$$T \propto \cos^2 \lambda\theta, \qquad (8.2.20)$$
$$c \propto \cos \lambda\theta.$$

Formulae (8.2.10)–(8.2.20) are useful in calculations. For example, suppose that, as illustrated in Figure 8.2.3, we are given the upstream Mach number M_1 and the wall deflection angle δ, and we wish to determine the downstream Mach number M_2 and the position of the fan, as specified by the angle μ_1 between its upstream edge and the extension of the upstream wall, and by the angle μ_2 between its downstream edge and the downstream wall. Since μ_1 and μ_2 are the upstream and downstream Mach angles (i.e., $\mu_1 = \sin^{-1}(1/M_1)$ and $\mu_2 =$

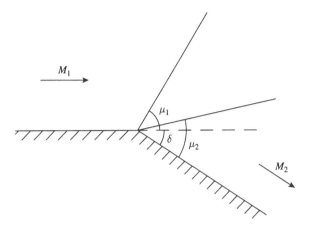

Fig. 8.2.3. Upstream and downstream Mach numbers M_1 and M_2, the corresponding Mach angles μ_1 and μ_2, and the deflection angle δ.

$\sin^{-1}(1/M_2))$ we obtain μ_1 at once, and we need to find M_2 as a function of M_1 and δ. We first obtain the position of the reference line $\theta = 0$ by noting that the upstream edge of the fan, whose position we have just found, is the line $\theta = \theta_1$ obtained from (8.2.14) by putting $M = M_1$. Therefore the reference-stream deflection angle at the upstream edge of the fan is $\nu_1 = \mu_1 + \theta_1 - \frac{1}{2}\pi$. Hence the reference-stream deflection angle at the downstream edge of the fan is $\nu_2 = \nu_1 + \delta = \mu_1 + \theta_1 + \delta - \frac{1}{2}\pi$, and M_2 may be determined from the Prandtl–Meyer function by solving the equation $\nu_2 = \nu(M_2)$ either numerically or with the aid of published tables. From M_2 we obtain μ_2, and we have thus determined the angles and the downstream Mach number in Figure 8.2.3. The pressure, density, temperature, and sound speed throughout the flow are determined by (8.2.20).

If the wall deflection δ is greater than the limiting flow deflection as $M_2 \to \infty$, that is, greater than the maximum flow deflection $\nu_\infty - \nu(M_1)$, then the fluid separates from the wall at the corner, to leave a fan-shaped vacuum region between the fluid and the downstream wall. This is illustrated in Figure 8.2.4 for $M_1 = 1$, in which case the upstream flow is that of the reference stream and $\nu_\infty - \nu(M_1)$ takes its greatest value ν_∞. In the figure, the deflection angle of the fluid is $(\{(\gamma + 1)/(\gamma - 1)\}^{1/2} - 1)\pi/2$, the angle of the expansion fan is $\{(\gamma + 1)/(\gamma - 1)\}^{1/2}\pi/2$, and the angle subtended by the fluid is $(\{(\gamma + 1)/(\gamma - 1)\}^{1/2} + 1)\pi/2$. For a diatomic gas, with $\gamma = 7/5$, these angles are $(\sqrt{6} - 1)\pi/2$, $\sqrt{6}\pi/2$, and $(\sqrt{6} + 1)\pi/2$, that is, $130.5°$, $220.5°$, and $310.5°$.

Whether a vacuum region is actually produced is determined by conditions downstream. If the fluid is discharging into a region of fluid in which the

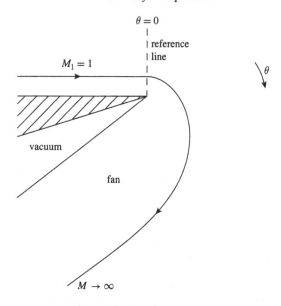

Fig. 8.2.4. A Prandtl–Meyer expansion of maximum angular extent.

pressure is not especially low, then, at wall deflection angles large enough to produce flow separation, a vortex sheet or slip line marks the boundary between separated flow and stationary fluid. We saw this pattern in Figure 7.3.3, in our description of an underexpanded jet emerging from a nozzle, and Figure 8.2.5 shows the flow pattern near the nozzle exit. The positions of the expansion fan and slip line attached to each corner are determined by the flow conditions upstream in the nozzle and by the condition that on the slip line the pressure equals the ambient pressure p_2.

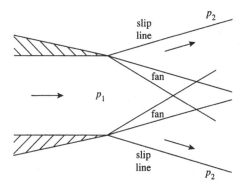

Fig. 8.2.5. Prandtl–Meyer expansion in an underexpanded jet produced by a nozzle with its exit pressure p_1 higher than the ambient pressure p_2.

8.3 Flow round a Smooth Bend

We shall see in Section 10.5 that supersonic flow round a smooth convex bend may be analysed by the method of characteristics. Alternatively, the flow may be regarded as built up from a sequence of small-angle Prandtl–Meyer expansion fans and analysed by the theory in the previous section. The two methods are related by the fact that the radial lines in an expansion fan are characteristics. If the smooth bend in Figure 8.3.1 is replaced by a polygon, the flow becomes a sequence of expansion fans attached to the vertices of the polygon, alternating with regions of rectilinear flow parallel to the sides of the polygon. In the limit as the length of the sides tends to zero, and their number tends to infinity, the expansion fans become the Mach lines of Figure 8.3.1.

More generally, if in a two-dimensional supersonic flow there is a region containing the Mach lines of a single family, and the pressure is decreasing downstream, the region is similar to a sequence of expansion fans. If the pressure is increasing downstream, the flow in the region resembles a sequence of "compression fans," obtained by reversing the flow direction in the analysis of the previous section. The boundary of a single-family region contains a shock wave if the Mach lines focus at a caustic, and it may also contain a line where the flow meets the Mach lines of another family. Both of these possibilities occur in the flow round the smooth concave bend shown in Figure 8.3.2. In the region to the left, the flow is a sequence of small-angle fans. The upper-right part of this region is bounded by the shock produced by the focusing of the Mach lines from the wall, and the lower-right part is bounded by the Mach line, of the other family, that begins at the lower end of the shock. The figure

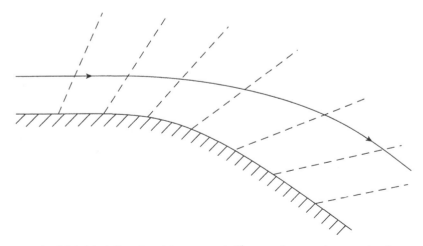

Fig. 8.3.1. Mach lines (– – –) in supersonic flow round a smooth convex bend.

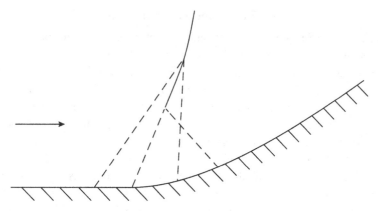

Fig. 8.3.2. Mach lines (– – –) and a shock (——) in supersonic flow round a smooth concave bend.

does not show the repeated reflections of Mach lines between the shock and the wall.

8.4 Bibliographic Notes

Prandtl–Meyer expansion was first described in Meyer (1908), which is based on work in Meyer's doctoral dissertation. The topic is treated in the reference articles by Bickley (1953) and Ferri (1955a) and in all textbooks on high speed flow.

8.5 Further Results and Exercises

1. A polytropic gas with ratio of specific heats γ flows inviscidly and super-sonically round a sharp corner and undergoes a Prandtl–Meyer expansion. Taking the corner as the origin of a polar coordinate system (r, θ), in which the velocity components are (u, v), show that one set of characteristics is the set of lines of constant θ and that, for some constants a and b, the other set of characteristics is given by $r^2 \propto u^a v^b$. Find the values of a and b in terms of γ, and deduce the equations of the streamlines.

2. In an expansive supersonic flow round a polygonal bend, a stream of poly-tropic gas with ratio of specific heats γ is deflected through the small angle θ_n at the nth corner ($n = 1, 2, 3, \ldots$), after which the pressure of the gas is p_n and the local Mach angle is μ_n. Prove that, approximately,

$$\frac{p_n}{p_{n-1}} = 1 - 2\gamma\theta_n \operatorname{cosec} 2\mu_n,$$

$$\mu_n = \mu_{n-1} + \theta_n\left(1 - \frac{1}{2}(\gamma + 1)\sec^2\mu_{n-1}\right).$$

Now assume that the bend is continuous, and denote the deflection of the gas by θ. Show that the pressure p and Mach angle μ satisfy

$$\frac{1}{p}\frac{dp}{d\theta} = -2\gamma \operatorname{cosec} 2\mu, \qquad \frac{d\mu}{d\theta} = 1 - \frac{1}{2}(\gamma + 1)\sec^2 \mu.$$

The undeflected stream has speed V_0 and Mach angle μ_0, and the deflected stream has speed V. The functions $f(\mu)$ and $g(\mu)$ are defined by

$$f(\mu) = -\left(\frac{\gamma+1}{\gamma-1}\right)^{\frac{1}{2}} \tan^{-1}\left\{\left(\frac{\gamma+1}{\gamma-1}\right)^{\frac{1}{2}} \tan \mu\right\} + \mu,$$

$$g(\mu) = \left(\frac{\sin^2 \mu}{\gamma - \cos 2\mu}\right)^{\frac{\gamma}{\gamma-1}}.$$

Show that

$$\theta = f(\mu) - f(\mu_0), \qquad \frac{p}{p_0} = \frac{g(\mu)}{g(\mu_0)}, \qquad \frac{V}{V_0} = \left(\frac{\gamma - \cos 2\mu_0}{\gamma - \cos 2\mu}\right)^{\frac{1}{2}}.$$

3. A uniform supersonic stream of a polytropic gas with ratio of specific heats γ expands around a sharp corner at the origin. During the expansion, the inclination of the direction of motion to the radius vector from the corner is μ, the angle the radius vector makes with a suitable initial line is θ, and the pressure is p. Define λ and n by $\lambda^2 = (\gamma-1)/(\gamma+1)$ and $n = 2\gamma/(\gamma-1)$, and let p_0 be a suitable reference pressure. Show that $\tan \mu = \lambda \cot \lambda\theta$ and $p = p_0 \cos^n \lambda\theta$.

4. A perfect gas is in steady two-dimensional motion. The ratio of specific heats of the gas is γ, and in polar coordinates (r, θ) the velocity components are (u, v). Show that, in terms of a speed $\hat{c}$, a reference angle θ_0, and the quantity λ defined by $\lambda^2 = (\gamma - 1)/(\gamma + 1)$, a possible motion of the gas is

$$u = \frac{\hat{c}}{\lambda} \sin \lambda(\theta - \theta_0), \qquad v = \hat{c} \cos \lambda(\theta - \theta_0).$$

Explain briefly how this solution can be applied to the expansion of a gas flowing round a convex corner. Show that, if the oncoming stream has Mach number 1, then the maximum angle through which the stream can be turned is $(\lambda^{-1} - 1)\pi/2$.

5. A uniform supersonic stream of a polytropic gas passes through a Prandtl–Meyer expansion fan as the gas expands round a sharp corner. Assume that, in standard notation, with suffices denoting partial differentiation, the

equations of motion describing the flow are

$$(\rho r u)_r + (\rho v)_\theta = 0,$$

$$uu_r + \frac{vu_\theta}{r} - \frac{v^2}{r} = -\frac{p_r}{\rho},$$

$$uv_r + \frac{vv_\theta}{r} + \frac{uv}{r} = -\frac{p_\theta}{\rho r}.$$

Determine the temperature at all positions in the flow. What happens if your formulae lead to temperatures of absolute zero?

9

Aerofoils

9.1 Linear Theory. Subsonic and Supersonic Flow

This chapter contains the elements of the theory of steady two-dimensional high speed flow past an aerofoil. In this section we present linear theory, in Section 9.2 we discuss nonlinear effects, and in Section 9.3 we obtain the scaling law and dominant nonlinear equation for transonic aerofoil theory.

We use a frame of reference in which the aerofoil is at rest and the undisturbed fluid is in uniform rectilinear motion. We ignore viscosity and thermal conductivity, so that our analysis applies outside of boundary layers, and we take the fluid to be a polytropic gas with ratio of specific heats γ, so that, for air, $\gamma = 7/5$. The flow is steady, two dimensional, irrotational, and homentropic. In Cartesian coordinates (x, y), the velocity of the fluid is $\mathbf{u} = (u, v) = (u(x, y), v(x, y))$, and, since the flow is irrotational, we may define a velocity potential ϕ such that $\mathbf{u} = \nabla\phi$, that is, $(u, v) = (\phi_x, \phi_y)$. When using suffix notation, we write $(x, y) = (x_1, x_2)$ and $(u, v) = (u_1, u_2)$. The fluid has pressure p and speed of sound c, and the undisturbed velocity, pressure, and speed of sound are $(U, 0)$, p_∞, and c_∞. The coordinates and definitions are illustrated in Figure 9.1.1.

The assumptions in the previous paragraph are reasonable for many purposes. For example, although the free stream will usually contain turbulent eddies, causing the aerofoil to be subject to unsteady loads and to radiate sound, the associated velocity perturbations are only a small proportion of the free stream velocity, and consequently they may be ignored when calculating the main force on the aerofoil. Similarly, the flow cannot be exactly homentropic if shocks are present, even ignoring viscosity and thermal conductivity, but we have seen that entropy changes are extremely small except at strong shocks. Thus the results we obtain in this section are useful in practice. Nevertheless, our assumptions have their limitations, as we shall see in Section 9.2 in our discussion of more complete theories.

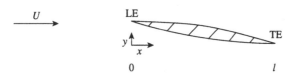

Fig. 9.1.1. Aerofoil of chord l in a fluid with velocity $\mathbf{u} = (u, v) = (\phi_x, \phi_y)$. The undisturbed free-stream velocity is $(u, v) = (U, 0)$. Leading edge, LE; trailing edge, TE.

The equations of motion for steady irrotational homentropic flow were given in Chapter 4, and the thermodynamic relations for a polytropic gas in Chapter 3. We need the steady flow equation (4.3.9), Bernoulli's equation (4.3.12) in the absence of body forces, and the relation between p and c at fixed entropy. Thus

$$c^2 \frac{\partial u_i}{\partial x_i} = u_i u_j \frac{\partial u_i}{\partial x_j}, \qquad (9.1.1a)$$

$$\frac{c^2}{\gamma - 1} + \frac{\mathbf{u}^2}{2} = \frac{c_\infty^2}{\gamma - 1} + \frac{U^2}{2}, \qquad (9.1.1b)$$

$$\frac{p}{p_\infty} = \left(\frac{c^2}{c_\infty^2}\right)^{\frac{\gamma}{\gamma-1}}. \qquad (9.1.1c)$$

The first equation is

$$(c^2 - u^2)u_x - uv(u_y + v_x) + (c^2 - v^2)v_y = 0, \qquad (9.1.2)$$

or

$$\left(c^2 - \phi_x^2\right)\phi_{xx} - 2\phi_x\phi_y\phi_{xy} + \left(c^2 - \phi_y^2\right)\phi_{yy} = 0. \qquad (9.1.3)$$

We now assume that the aerofoil is thin and lightly cambered (i.e., has no sudden changes in the slope of its surfaces) and is at a small angle of attack to the oncoming stream. Then the aerofoil produces only a small perturbation to the free stream, and we may perform a linear perturbation analysis about the solution of the equations when the aerofoil is absent. Since the undisturbed potential is Ux, we define the perturbation potential $\hat{\phi}(x, y)$ by

$$\phi = Ux + \hat{\phi}. \qquad (9.1.4)$$

Thus

$$(\phi_x, \phi_y) = (U + \hat{\phi}_x, \hat{\phi}_y). \tag{9.1.5}$$

Substitution into (9.1.3) gives an equation for $\hat{\phi}$, which on linearisation (i.e., with products of $\hat{\phi}$ and its derivatives ignored) is

$$\left(c_\infty^2 - U^2\right)\hat{\phi}_{xx} + c_\infty^2 \hat{\phi}_{yy} = 0. \tag{9.1.6}$$

The Mach number M of the flow is defined by

$$M = \frac{U}{c_\infty}. \tag{9.1.7}$$

Thus M is a constant, and

$$(1 - M^2)\hat{\phi}_{xx} + \hat{\phi}_{yy} = 0. \tag{9.1.8}$$

Bernoulli's equation (9.1.1b) gives c^2 in terms of $\hat{\phi}$ as

$$\frac{c^2}{c_\infty^2} = 1 + \frac{\gamma - 1}{2c_\infty^2}\left\{U^2 - (U + \hat{\phi}_x)^2 - \hat{\phi}_y^2\right\}. \tag{9.1.9}$$

Linearised, this is

$$\frac{c^2}{c_\infty^2} \simeq 1 - (\gamma - 1)\frac{U\hat{\phi}_x}{c_\infty^2}. \tag{9.1.10}$$

The relation between p and c^2 in (9.1.1c) gives

$$\frac{p}{p_\infty} \simeq 1 - \frac{\gamma U\hat{\phi}_x}{c_\infty^2}. \tag{9.1.11}$$

In terms of the undisturbed fluid density ρ_∞, which satisfies $c_\infty^2 = \gamma p_\infty/\rho_\infty$, we have

$$p - p_\infty \simeq -\rho_\infty U\hat{\phi}_x. \tag{9.1.12}$$

Let the upper and lower surfaces of the aerofoil be at $y = f_+(x)$ and $y = f_-(x)$. By assumption, the slopes $f'_+(x)$ and $f'_-(x)$ are small in absolute value. In linear theory, the oncoming stream acquires at the surfaces a y component of velocity $Uf'_+(x)$ and $Uf'_-(x)$, and the resulting boundary condition, which because of the thinness of the aerofoil may be applied at $y = 0^+$ and $y = 0^-$ instead of exactly on the surfaces, is

$$\hat{\phi}_y = \begin{cases} Uf'_+(x) & (y = 0^+), \\ Uf'_-(x) & (y = 0^-). \end{cases} \qquad (9.1.13)$$

We also impose the radiation condition that at infinity any waves are outgoing. The solution of the full nonlinear problem has a stagnation point near the leading edge of the aerofoil, so that, in a small region nearby, the perturbation of the undisturbed fluid is large, and linear theory cannot be accurate. A full analysis shows that this does not invalidate the linear theory in the rest of the flow.

When the flow is subsonic (i.e., $M < 1$) Equation (9.1.8) is elliptic, because the coefficient $1 - M^2$ of $\hat{\phi}_{xx}$ is positive. Suitable scalings of x, y, the boundary conditions, and the dependent variables lead to a boundary-value problem for Laplace's equation, equivalent to a problem of incompressible flow in the scaled variables. Therefore under the assumptions we have made, the theory of subsonic flow past an aerofoil is reducible to the theory of incompressible flow, on which there is an extensive mathematical literature, including techniques based on, for example, complex variables, Hilbert transforms, eigenfunctions, and the application of the Kutta condition at the trailing edge. As these techniques are thoroughly covered in texts on incompressible flow, we shall not discuss them here. The scalings that reduce a subsonic problem to an incompressible problem are called the Prandtl–Glauert correction factors for compressibility.

When the flow is supersonic (i.e., $M > 1$) Equation (9.1.8) is hyperbolic, because $1 - M^2$ is negative. Let us define the positive quantity β by

$$\beta^2 = M^2 - 1. \qquad (9.1.14)$$

Then (9.1.8) becomes the wave equation

$$\beta^2 \hat{\phi}_{xx} = \hat{\phi}_{yy}. \qquad (9.1.15)$$

An interpretation of (9.1.5), as used to describe the waves indicated in Figure 9.1.2, is that x is a scaled timelike variable and y is a space variable, and the straight lines in the figure are propagation paths away from a source on $y = 0$. Since the wave equation governs the propagation of sound waves, it follows that linear supersonic aerofoil theory and linear acoustic theory are really the same subject. For this reason, the linear theory of supersonic aerodynamics is often

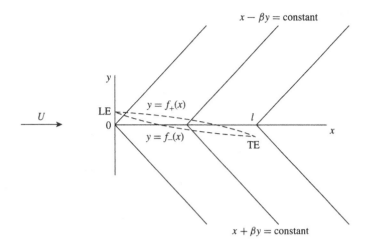

Fig. 9.1.2. Aerofoil with upper and lower surfaces $y = f_+(x)$ and $y = f_-(x)$. In the thin aerofoil approximation the boundary conditions are applied on $y = 0^+$ and $y = 0^-$. The lines $x \pm \beta y = \text{constant}$ are disturbance paths.

called the acoustic version of aerodynamics. Shocks in linear acoustic theory arise as discontinuities across characteristic surfaces, and these are sometimes called sound pulses.

The general solution of (9.1.15) may be written in terms of two arbitrary functions f and g as

$$\hat{\phi} = f(x + \beta y) + g(x - \beta y). \tag{9.1.16}$$

For $y > 0$, the radiation condition that waves are outgoing implies that $\hat{\phi}$ can only contain g [i.e., that $\hat{\phi} = g(x - \beta y)$]. For $y = 0^+$, the boundary condition (9.1.13) gives $-\beta g'(x) = U f_+'(x)$. We take $g(x) = -(U/\beta) f_+(x)$, ignoring a constant of integration, which does not affect the pressure or velocity, so that $g(x - \beta y) = -(U/\beta) f_+(x - \beta y)$. Similarly, for $y < 0$ we find that $\hat{\phi}$ can only contain f and that $f(x) = (U/\beta) f_-(x)$, so that $f(x + \beta y) = (U/\beta) f_-(x + \beta y)$. Therefore the solution of (9.1.15) that satisfies the required conditions is

$$\hat{\phi} = \begin{cases} \dfrac{-U}{\beta} f_+(x - \beta y) & (y > 0), \\[2mm] \dfrac{U}{\beta} f_-(x + \beta y) & (y < 0). \end{cases} \tag{9.1.17}$$

Then (9.1.12) gives

$$
p - p_\infty = \begin{cases} \dfrac{\rho_\infty U^2}{\beta} f'_+(x) & (y = 0^+), \\[3mm] \dfrac{-\rho_\infty U^2}{\beta} f'_-(x) & (y = 0^-). \end{cases} \tag{9.1.18}
$$

Therefore on the aerofoil surface, the pressure change $p - p_\infty$ from the ambient pressure p_∞ is proportional to the slope of the surface. In (9.1.17) and (9.1.18), and subsequent formulae, the denominator contains the factor $\beta = (M^2 - 1)^{\frac{1}{2}}$, which tends to zero when $M \to 1$. Thus linear theory needs to be revised when M is close to 1 (i.e., in the transonic regime). The necessary revision is presented in Section 9.3. The lines $x \pm \beta y = $ constant, on which ϕ and p are constant, are shown in Figure 9.1.2.

Assume that the aerofoil, of chord l, has its leading edge at $(0, y_{\mathrm{LE}})$ and its trailing edge at (l, y_{TE}). By the assumptions of linear theory, the camber is small and the angle of attack α satisfies $\alpha \simeq (y_{\mathrm{LE}} - y_{\mathrm{TE}})/l \ll 1$. The force on the aerofoil in the direction perpendicular to the free stream, per unit length parallel to the leading or trailing edge (i.e., per unit length in the span direction) is the lift L. The value of L, which is positive when the lift is upwards, is given by a line integral $\oint p\,dx$, anticlockwise around the aerofoil, of the surface pressure in (9.1.18). Thus with proper allowance for signs,

$$
\begin{aligned}
L &= \oint p\,dx \\
&= \frac{-\rho_\infty U^2}{\beta} \int_0^l \{f'_-(x) + f'_+(x)\}\,dx \\
&= \frac{2\rho_\infty U^2}{\beta}(y_{\mathrm{LE}} - y_{\mathrm{TE}}) \\
&= \frac{2\rho_\infty U^2 l\alpha}{\beta}.
\end{aligned} \tag{9.1.19}
$$

The lift force, per unit area of the aerofoil and normalised by $\frac{1}{2}\rho_\infty U^2$, is the lift coefficient C_{L}. Thus

$$
C_{\mathrm{L}} = \frac{L/l}{\frac{1}{2}\rho_\infty U^2} = \frac{4}{\beta}\left(\frac{y_{\mathrm{LE}} - y_{\mathrm{TE}}}{l}\right) = \frac{4\alpha}{\beta}. \tag{9.1.20}
$$

Therefore according to linear theory, the lift does not depend on the detailed shape of the aerofoil, but only on the angle of attack and the Mach number.

The force on the aerofoil in the free-stream direction, per unit length in the span direction, is the drag D. Our simplified theory does not determine the contribution to D made by tangential forces due to viscosity (i.e., by skin friction), and we shall calculate only the contribution made by p in (9.1.18). This is the wave drag D_{wave}, equal to the line integral $\oint p\,dy$ around the aerofoil. Thus

$$
\begin{aligned}
D_{\text{wave}} &= \oint p\,dy \\
&= \oint p\frac{dy}{dx}\,dx \\
&= \frac{\rho_\infty U^2}{\beta}\int_0^l \left\{f_-'^2(x) + f_+'^2(x)\right\}dx.
\end{aligned}
\tag{9.1.21}
$$

Let the slopes of the upper and lower surfaces deviate from α by $\theta_+(x)$ and $\theta_-(x)$, so that

$$
\begin{aligned}
f_+'(x) &= -\alpha + \theta_+(x), \\
f_-'(x) &= -\alpha + \theta_-(x).
\end{aligned}
\tag{9.1.22}
$$

The mean values of $\theta_+^2(x)$ and $\theta_-^2(x)$ over the chord will be denoted $\langle\theta_+^2\rangle$ and $\langle\theta_-^2\rangle$. Then (9.1.21) gives

$$
D_{\text{wave}} = \frac{2\rho_\infty U^2 l}{\beta}\left\{\alpha^2 + \frac{1}{2}\left(\langle\theta_+^2\rangle + \langle\theta_-^2\rangle\right)\right\}.
\tag{9.1.23}
$$

The wave-drag force, per unit area of the aerofoil and normalised by $\frac{1}{2}\rho_\infty U^2$, is the wave-drag coefficient $C_{\text{D,wave}}$. Thus

$$
\begin{aligned}
C_{\text{D,wave}} &= \frac{D_{\text{wave}}/l}{\frac{1}{2}\rho_\infty U^2} \\
&= \frac{4}{\beta}\left\{\alpha^2 + \frac{1}{2}\left(\langle\theta_+^2\rangle + \langle\theta_-^2\rangle\right)\right\}.
\end{aligned}
\tag{9.1.24}
$$

The contribution $4\alpha^2/\beta$ depends only on the angle of attack and the Mach number and is the same as for a flat plate at angle of attack α. The other contribution, $(2/\beta)(\langle\theta_+^2\rangle + \langle\theta_-^2\rangle)$, measures the effect on the aerofoil of the waves produced in the fluid by the nonplanar shape of the aerofoil surface. Since the radiation of wave energy is always away from the aerofoil, the effect is always to increase the drag. Some examples of formulae (9.1.18)–(9.1.24) applied to particular aerofoil shapes are given in the further results and exercises in Section 9.5.

9.2 Nonlinear Theory

The linear theory presented in the previous section is not always successful, most notably (a) at large distances from the aerofoil, that is, in the far field; (b) at high Mach numbers, that is, in the hypersonic regime; and (c) at Mach numbers close to unity, that is, in the transonic regime. We shall discuss in turn these three cases, which apply at the appropriate distances and Mach numbers no matter how thin the aerofoil.

During propagation in air, an acoustic wave gradually distorts from the constant shape predicted by linear theory, because high pressures travel a little faster than low pressures. Thus the wave steepens where pressure is decreasing with distance, and flattens where it is increasing, until the wave is N-shaped, with each vertical part of the N consisting of a shock. At still greater distances, the shocks thicken and lose their identity as the wave becomes smooth again. The theory of nonlinear wave steepening, shock formation, and decay is extensive and important, particularly in studies of sonic boom, and draws heavily on work by Whitham, leading to his F-function, and on the method of matched asymptotic expansions.

At high Mach number, the change in pressure caused by the motion of the aerofoil is a significant fraction of the original pressure p_∞, and nonlinear terms in the equations of motion become important. Even a thin aerofoil then produces strong shocks, and the consequent production of entropy gradients and vorticity invalidates the original assumption of irrotational homentropic flow. Nevertheless, the results of the previous section correctly determine, as follows, the dimensionless parameter required for aerofoil theory at high Mach number.

We let the functions $f'_\pm(x)$, which measure the slopes of the aerofoil surfaces, have order of magnitude θ. Then $\theta \ll 1$, because the aerofoil is thin and at a small angle of attack, and θ is also the order of magnitude of the angular deflection of the flow near the aerofoil. Equation (9.1.18) for $p - p_\infty$, together with $\beta = (M^2 - 1)^{\frac{1}{2}}$, $p_\infty = \gamma^{-1} \rho_\infty c_\infty^2$, and $M \gg 1$, gives $(p - p_\infty)/p_\infty \simeq \gamma M \theta$. Therefore nonlinear theory is needed when $M\theta \geq O(1)$ (i.e., $M \geq O(1/\theta)$). This condition defines the hypersonic regime, and $M\theta$ is the hypersonic similarity parameter.

An aerofoil at Mach number close to unity is travelling at approximately the same speed as the pressure waves it produces (i.e., the speed of sound). Therefore reinforcement – the superposition of pressure waves with similar phases – leads to high pressures near the aerofoil. The insufficiency of linear theory when $M \to 1$ is indicated by the factor $(M^2 - 1)^{\frac{1}{2}}$ in denominators. We shall obtain in the next section the dimensionless parameter, called the transonic similarity parameter, that determines the transonic regime of aerofoil theory.

Each of the three areas in which linear theory is insufficient – in the far field, in the hypersonic regime, and in the transonic regime – leads, on further analysis, to a new body of theoretical results, each of great practical importance. Several books are devoted exclusively to hypersonic flow or to transonic flow. In this book we present further analysis only for the transonic regime.

9.3 Transonic Flow

This section begins with a summary of the main experimental facts concerning steady two-dimensional transonic flow past an aerofoil. It continues with the mathematical theory leading to the transonic similarity parameter and the transonic small disturbance equation.

At a given free-stream Mach number M_∞ (i.e., Mach number at infinity), the fluid around a stationary aerofoil may contain regions of subsonic flow and regions of supersonic flow. A boundary between them may be shock free, in which case it is a line on which the Mach number equals 1 (i.e., a sonic line), or it may consist of a shock, separating an upstream supersonic region from a downstream subsonic region. A shock may also occur entirely within a supersonic region. In diagrams representing flow patterns, we indicate a subsonic region by $-$, a supersonic region by $+$, a sonic line by a dashed line, and a shock by a solid line. The coordinates are x in the free-stream direction and y in the transverse direction.

Consider the changes in the flow pattern as M_∞ is gradually increased through the subsonic, transonic, and supersonic regimes. The first value of M_∞ at which the flow is anywhere sonic or supersonic is the critical Mach number M_c and is well below 1. Figures 9.3.1a–g show the flow patterns for (a) $M_\infty < M_c$; (b) $M_c < M_\infty < 1$; (c) $M_\infty = 1^-$; (d) $M_\infty = 1$; (e) $M_\infty = 1^+$; (f) $M_\infty > 1$, with M_∞ fairly close to 1; and (g) $M_\infty > 1$, with M_∞ not too close to 1 and not too large. We describe the patterns in turn.

(a) The Regime $M_\infty < M_c$

The flow is everywhere subsonic, and there are no sonic lines or shocks.

(b) The Regime $M_c < M_\infty < 1$

There is at least one bounded supersonic region. Such a "supersonic bubble" is bounded in front by a sonic line, at the rear mostly by a shock, and on one side by the aerofoil surface.

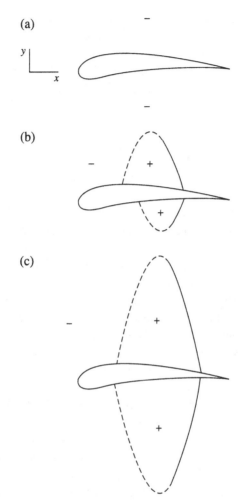

Fig. 9.3.1. Steady flow past an aerofoil as the free-stream Mach number M_∞ is increased through the critical Mach number M_c and through the sonic Mach number $M_\infty = 1$. Subsonic regions ($-$) are separated from supersonic regions ($+$) by sonic lines ($---$) or shocks (———). The stream coordinate is x, and the transverse coordinate is y. (a) $M_\infty < M_c$. Subsonic everywhere; no shocks. (b) $M_c < M_\infty < 1$. Supersonic bubbles. (c) $M_\infty = 1^-$. Supersonic bubbles elongate in transverse direction, and shocks move backwards. (d) $M_\infty = 1$. Supersonic bubbles extend to infinity. (e) $M_\infty = 1^+$. Bow shock ahead of leading edge; tail shocks attached to trailing edge. (f) $M_\infty > 1$, with M_∞ fairly close to 1. Compared with (e), bow shock is closer to leading edge, and subsonic bubble is smaller. (g) $M_\infty > 1$, with M_∞ not too close to 1 and not too large. Subsonic bubble now very small. Flow well described by linear theory, except near nose and in far field.

(d)

(e)

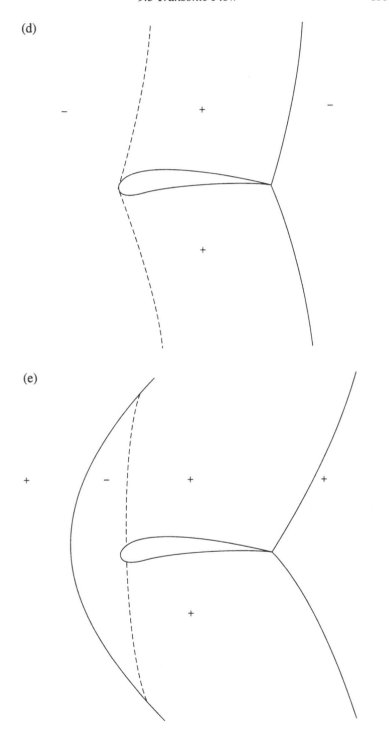

Fig. 9.3.1. (*Continued*)

(f)

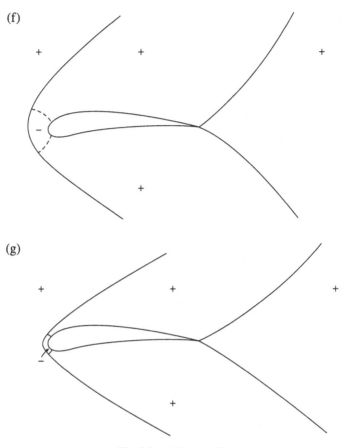

(g)

Fig. 9.3.1. (*Continued*)

(c) The Regime $M_\infty = 1^-$

The shocks marking the rear of the supersonic bubbles have moved backwards and meet the aerofoil near the trailing edge. The supersonic bubbles are elongated in the transverse direction.

(d) The Regime $M_\infty = 1$

The free-stream velocity is exactly sonic. The shocks meet the aerofoil surface at the trailing edge (i.e., they are tail shocks) and the supersonic bubbles, now elongated to infinity, tilt in the free-stream direction in accordance with the far-field scaling $y \propto x^{\frac{5}{4}}$, obtainable from a similarity solution of the transonic equations. The sonic lines and the shocks extend to infinity.

(e) The Regime $M_\infty = 1^+$

A bow shock extends to infinity and is met by a sonic line from the aerofoil, so that ahead of the aerofoil is a leading edge subsonic bubble. Thus the flow behind the bow shock is subsonic near the aerofoil and supersonic further away. The tail shocks have supersonic flow on each side.

(f) The Regime $M_\infty > 1$, with M_∞ fairly close to 1

The bow shock has moved closer to the leading edge, and the leading edge subsonic bubble has become smaller.

(g) The Regime $M_\infty > 1$, with M_∞ not too close to 1 and not too large

The leading edge subsonic bubble is now very small, so that the flow is almost everywhere supersonic. Except at large distances or at large M_∞, the flow is well described by linear theory. Thus Figure 9.3.1g is closely related to Figure 9.1.2.

We have not displayed the coalescing pairs of shocks that form next to the aerofoil surface, the so-called λ-shocks on account of their distinctive shape, nor have we discussed separation of the boundary layer near the region where a shock intersects it (i.e., shock-induced boundary-layer separation) or the dependence of the flow pattern on the shape of the aerofoil and its angle of attack. Thus Figures 9.3.1a–g are a little simplified. Nevertheless, they show many important features of transonic flow past an aerofoil.

To derive the transonic equation, we assume that the aerofoil is thin enough for all shocks produced to be weak, so that the flow is very nearly irrotational and homentropic, by (6.2.19) and Crocco's equation (4.3.4). Thus with only small error we may put $\mathbf{u} = \nabla\phi$. We shall exploit the fact, noted in our comments on Figures 9.3.1c–e, that when the speed of the free stream is close to sonic, the transverse length scale of variation is much greater than that in the stream direction. This is apparent from the linear equation (9.1.8), which requires a balance between $(1 - M^2)\partial^2/\partial x^2$ and $\partial^2/\partial y^2$, so that M close to 1 gives $\partial/\partial y \ll \partial/\partial x$.

Let us return to the full nonlinear equation (9.1.3) for the potential ϕ and concentrate on the term $(c^2 - \phi_x^2)\phi_{xx}$. The factor $c^2 - \phi_x^2$ becomes small at transonic speeds and can change sign within the fluid. Recall that the perturbation potential $\hat{\phi}$ is defined by $\phi = Ux + \hat{\phi}$, so that from Bernoulli's equation (9.1.9) we have

$$c^2 - \phi_x^2 = c_\infty^2 - U^2 - (\gamma + 1)U\hat{\phi}_x - \tfrac{1}{2}(\gamma + 1)\hat{\phi}_x^2 - \tfrac{1}{2}(\gamma - 1)\hat{\phi}_y^2. \quad (9.3.1)$$

Since the aerofoil is thin, we may neglect the quadratic terms on the right of
(9.3.1). To obtain an approximation to (9.1.3), with $M_\infty = U/c_\infty$ close to 1,
and with $\partial/\partial y \ll \partial/\partial x$, we keep the nonlinear term arising from $(\gamma + 1)U\hat{\phi}_x$,
to allow the sign of $c^2 - \phi_x^2$ to vary in the fluid, but we ignore all other nonlinear
terms. Thus approximately

$$\left(1 - M_\infty^2 - \frac{(\gamma + 1)U\hat{\phi}_x}{c_\infty^2}\right)\hat{\phi}_{xx} + \hat{\phi}_{yy} = 0. \qquad (9.3.2)$$

We shall shortly check the validity of (9.3.2) by a formal scaling analysis. The
upper and lower surfaces of the aerofoil, at $y = f_+(x)$ and $y = f_-(x)$, give
approximate boundary conditions $\hat{\phi}_y = Uf'_\pm(x)$ on $y = 0^\pm$.

We now perform a length-scale analysis of transonic flow past a thin aerofoil, to show that the terms in (9.3.2) have the same order of magnitude and are
larger than the terms neglected from (9.1.3). This will establish the validity of
(9.3.2) as an approximate equation worthy of study. Let the aerofoil have chord
l and occupy the region $0 < x < l$. The maximum thickness of the aerofoil is a
proportion δ of the chord (i.e., δl), and we assume that $\delta \ll 1$. The positions of
the upper and lower surfaces of the aerofoil will be specified in terms of dimensionless order-one functions F_+ and F_- by $y/(\delta l) = F_\pm(x/l)$. Figures 9.3.1c–e
for the transonic regime suggest that the length scale l of variation in the stream
direction applies not only close to the aerofoil but also well into the fluid, so
that the appropriate dimensionless stream coordinate is $x' = x/l$. In the transverse direction, the length scale, though clearly larger than l, is not apparent.
Therefore we introduce a small dimensionless parameter η defined so that this
length scale is l/η and the appropriate dimensionless transverse coordinate
is $y' = \eta y/l$. Thus in terms of x' and y', the aerofoil surfaces are at $y' =
\eta\delta F_\pm(x')$.

Since the perturbation potential $\hat{\phi}$ has the dimensions of velocity multiplied
by length (i.e., $U l$), we introduce a small dimensionless parameter ϵ so that the
scale of $\hat{\phi}$ is $\epsilon U l$. Then the length scales l and l/η in the x and y directions allow
us to define a dimensionless order-one potential ϕ' by $\hat{\phi}/(\epsilon U l) = \phi'(x/l, \eta y/l)$,
so that $\hat{\phi} = \epsilon U l\phi'(x', y')$.

The slopes of the upper and lower surfaces of the aerofoil are $dy/dx =
\delta F'_\pm(x/l) = \delta F'_\pm(x')$, and the transverse component of velocity is $\hat{\phi}_y =
\epsilon\eta U\phi'_{y'}(x', y')$. Therefore on $y = 0^\pm$ the boundary condition $\hat{\phi}_y = Udy/dx$
becomes $\epsilon\eta U\phi'_{y'} = U\delta F'_\pm$ [i.e., $\phi'_{y'}(x', 0^\pm) = (\delta/(\epsilon\eta))F'_\pm(x')$]. Since $\phi'_{y'}$ and
$F'_\pm$ are of order one, so also is $\delta/(\epsilon\eta)$, and by suitable definition of ϵ or η its
value may be taken to be exactly 1. Thus

$$\epsilon\eta = \delta. \qquad (9.3.3)$$

The scales of the terms in (9.3.2) are

$$\hat{\phi}_{xx} = \frac{\epsilon U}{l}\phi'_{x'x'},$$

$$\hat{\phi}_{yy} = \frac{\epsilon \eta^2 U}{l}\phi'_{y'y'}, \tag{9.3.4}$$

$$\frac{U\hat{\phi}_x}{c_\infty^2} = \frac{\epsilon U^2}{c_\infty^2}\phi'_{x'}.$$

Therefore

$$\left(1 - M_\infty^2 - \epsilon(\gamma + 1)M_\infty^2\phi'_{x'}\right)\phi'_{x'x'} + \eta^2\phi'_{y'y'} = 0. \tag{9.3.5}$$

To balance the terms in this equation we must ensure that the coefficients $1 - M_\infty^2$, ϵ, and η^2 have the same order of magnitude. The definitions of ϵ and η still have enough flexibility for two of the three coefficients to be taken exactly equal, and so we may assume that

$$\epsilon = \eta^2. \tag{9.3.6}$$

The relations (9.3.3) and (9.3.6) together imply that

$$\epsilon = \delta^{\frac{2}{3}}, \qquad \eta = \delta^{\frac{1}{3}}. \tag{9.3.7}$$

Therefore $1 - M_\infty^2$ must be of order $\delta^{\frac{2}{3}}$. Accordingly, we define an order-one quantity K, called the transonic similarity parameter, by

$$K = \frac{1 - M_\infty^2}{\delta^{\frac{2}{3}}}. \tag{9.3.8}$$

The transonic similarity parameter may be positive, zero, or negative.

The earlier definitions and scalings may be rewritten using only l, δ, and K. Recall that δ is the thickness-to-chord ratio of the aerofoil. Thus the upper and lower surfaces of the aerofoil and the scalings of x, y, ϕ, and M_∞^2 are

$$y = \delta l F_\pm(x/l),$$

$$(x, y) = (lx', \delta^{-\frac{1}{3}}ly'),$$

$$\phi = Ux + \delta^{\frac{2}{3}}Ul\phi', \tag{9.3.9}$$

$$M_\infty^2 = 1 - K\delta^{\frac{2}{3}}.$$

To leading order, (9.3.5) is

$$(K - (\gamma + 1)\phi'_{x'})\phi'_{x'x'} + \phi'_{y'y'} = 0. \tag{9.3.10}$$

A check reveals that, with the scalings (9.3.9), the terms omitted from the full Equations (9.1.1) or (9.1.3) are proportional to higher powers of δ than are the terms leading to (9.3.10) and that only a negligible error, as measured by powers of δ, is introduced by applying the aerofoil boundary condition on the line $y = 0$ instead of on the aerofoil surface. Therefore (9.3.10) is a valid leading-order equation. It has been called the Kármán–Guderley equation, but it is now usually called the transonic small disturbance equation. A consequence of the scalings (9.3.9) is that, in the transonic regime, the pressure change due to an aerofoil scales with $\delta^{\frac{2}{3}}$, that is, the nondimensional pressure perturbation $(p - p_\infty)/(\rho_\infty U^2)$ is of order $\delta^{\frac{2}{3}}$.

An interpretation of the transonic small disturbance equation is that it consists of the dominant terms in the full equation for ϕ when the free-stream Mach number M_∞ tends to 1, either from above or below, and simultaneously the thickness-to-chord ratio of the aerofoil decreases in proportion to $|1 - M_\infty^2|^{\frac{3}{2}}$, so that K is fixed. The coefficient $K - (\gamma + 1)\phi'_{x'}$ of $\phi'_{x'x'}$, at fixed K, may be positive, zero, or negative for different (x', y'), depending on the values of $\phi'_{x'}$. Thus the equation is of mixed type: elliptic in some regions of space and hyperbolic in others. It has been extensively studied, especially by numerical methods using various techniques of upwind differencing and numerical shock-capturing to overcome the difficulty that the locations of the elliptic regions and hyperbolic regions are not known in advance. Exact solutions of the equation have also been found and studied, for example of the type described in Question 10 in Section 9.5. The transonic small disturbance equation reproduces the flow features shown in Figure 9.3.1 and provides an outstandingly successful model for many aspects of transonic flow.

9.4 Bibliographic Notes

Most topics in high speed flow are relevant to some aspect of flow past an aerofoil, and the following notes cover only a selection of items from the bibliography. Early papers on aerofoils in Table 13.2.1 are Glauert (1928), Küssner (1941), and Lighthill (1944). Reference works on aerofoils in Table 13.3.1 are Mair & Beavan (1953), Frick (1957), Garrick (1957), and Jones & Cohen (1957).

Aspects of aerofoil theory covered in the texts and monographs in Table 13.4.1 are subsonic and supersonic flow in Ward (1955), supersonic flow in Miles (1959), and transonic flow in Landahl (1961), Ferrari & Tricomi (1968), and Cole & Cook (1986). Aspects of transonic aerofoil theory covered in the surveys and reviews in Table 13.5.1 are aerofoil design in Nieuwland & Spee (1973) and Sobieczky & Seebass (1984), viscosity in Ryzhov (1978), aerofoil oscillations in Tijdeman & Seebass (1980), and numerical computation in Caughey (1982). The mathematical theory of transonic flow, some of it abstruse, is discussed in Lighthill (1947), Germain & Bader (1952), Morawetz

(1956, 1957), and Bers (1958). Three publications that greatly influenced numerical computation of transonic flows were Murman & Cole (1971), Bauer, Garabedian & Korn (1972), and Jameson (1974).

Aspects of aerofoil theory investigated in the research papers in Table 13.6.1d include topics in unsteady aerodynamics, for example gust interactions and aeroacoustics, analysed in Peake (1992a,b, 1993, 1994a,b), Amiet (1993, 1995), Myers & Kerschen (1995, 1997), Leppington & Sisson (1997), and Peake & Kerschen (1997); transonic flow in Rusak (1993); the delta wing in Chang & Lei (1996); and aerofoils in a fluid that allows rarefaction shocks in Monaco, Cramer & Watson (1997) and Rusak & Wang (1997).

9.5 Further Results and Exercises

1. A symmetrical two-dimensional aerofoil, of chord c and small nondimensional thickness of order ϵ, has its upper and lower surfaces at $y = \pm\epsilon x(1 - x/c)^2$. The aerofoil is at zero incidence in a steady supersonic stream flowing in the positive x direction at Mach number M. Assuming the validity of linear theory, find the velocity components in the upper region of disturbance, and show that the aerofoil has drag coefficient $(8/15)(M^2 - 1)^{-\frac{1}{2}}\epsilon^2$.

2. A two-dimensional flat-plate aerofoil of chord c is at zero incidence in a supersonic stream flowing in the positive x direction at Mach number M and oscillates transversely to the stream, without rotation, in the (x, y) plane at frequency ω and with dimensionless amplitude ϵ. The displacement of the aerofoil from $y = 0$ at time t is $\epsilon c \cos \omega t$. The incident stream has velocity U_∞, and the perturbation to the stream has velocity potential $U_\infty \phi$. Show that ϕ satisfies

$$M_\infty^2 \left(\frac{1}{U_\infty} \frac{\partial}{\partial t} + \frac{\partial}{\partial x} \right)^2 \phi = \frac{\partial^2 \phi}{\partial x^2} + \frac{\partial^2 \phi}{\partial y^2}.$$

Show that, with $B = (M^2 - 1)^{\frac{1}{2}}$ and a suitable choice of a constant k, the substitution $\phi = \exp\{-i(\omega t + kx)\} f(x, y)$ leads to the equation

$$B^2 \frac{\partial^2 f}{\partial x^2} - \frac{\partial^2 f}{\partial y^2} = -\left(\frac{B^2 + 1}{B^2} \right) \left(\frac{\omega}{U_\infty} \right)^2 f.$$

Solve the equation for ϕ when ω is sufficiently small that $(\omega c / U_\infty)^2$ is negligible, and hence determine the lift coefficient of the aerofoil.

3. Show that the wave equation $u_{yy} + u_{zz} = B^2 u_{xx}$ may be reduced by the transformation $By/x = r \cos \theta$, $Bz/x = r \sin \theta$ to a form in which u is a

function only of y/x and z/x and satisfies the equation

$$r^2(r^2 - 1)u_{rr} + r(2r^2 - 1)u_r - u_{\theta\theta} = 0.$$

Show that this equation may be reduced by a real transformation to $u_{\sigma\sigma} = u_{\theta\theta}$ if $r > 1$, and to $u_{ss} + u_{\theta\theta} = 0$ if $r < 1$. How is this relevant to linearised supersonic flow about a symmetrical thin delta wing with subsonic or supersonic leading edges?

4. A uniform supersonic stream of air is disturbed by a thin two-dimensional aerofoil at a small angle of attack α. Use linear theory to calculate the lift and wave drag coefficients of the aerofoil, and show that their ratio may be written in terms of a certain function k of the shape of the aerofoil as

$$\frac{L}{D} = \frac{1}{\alpha + k/\alpha}.$$

Give an expression for k.

   A thin aerofoil of rhombus-shaped cross section has chord c and thickness t. Determine the highest value of L/D obtainable from the aerofoil. When this value is attained, what are the angles between the sides of the rhombus and the oncoming stream?

5. A thin cylinder, lying approximately in the plane $y = 0$ and with its axis parallel to the z axis, is surrounded by a compressible fluid that may be assumed to have no viscosity or conductivity and at infinity has uniform velocity parallel to the x axis. The fluid has pressure p, density ρ, and velocity $\mathbf{u}$, and at infinity it has pressure p_∞, density ρ_∞, speed U_∞, and Mach number M_∞. Define the dimensionless pressure p', dimensionless density ρ', and scaled velocity potential ϕ by $p = p_\infty(1 + p')$, $\rho = \rho_\infty(1 + \rho')$, and

$$\mathbf{u} = (U_\infty(1 + \phi_x), U_\infty\phi_y, 0).$$

Write down the equations of motion of the fluid. Show that they reduce, when the disturbance to the flow is small enough that squares and products of p', ρ', and ϕ may be neglected, to

$$\left(1 - M_\infty^2\right)\phi_{xx} + \phi_{yy} = 0.$$

What are the boundary conditions satisfied by ϕ? Explain briefly the qualitative features of the flow when $M_\infty < 1$ and when $M_\infty > 1$, and illustrate your answer with sketches of the streamlines of the flow.

6. Explain the essential points of the linearised theory of the flow past a thin two-dimensional aerofoil with a sharp trailing edge.

   A two-dimensional aerofoil, lying close to the x axis between $x = -2a$ and $x = 2a$ and of dimensionless thickness of order $\epsilon \ll 1$, has its lower surface flat on the x axis and its upper surface at

$$y = \epsilon a \left(1 - \frac{x^2}{4a^2} \right).$$

   The aerofoil is placed in a stream that flows at a small angle α to the x axis. Show that under certain assumptions, which should be stated, the lift coefficient C_L is approximately

$$C_L = 2\pi \left(\alpha + \tfrac{1}{4}\epsilon \right).$$

7. Air flows in the (x, y) plane at high speed over an infinite corrugated surface $y = a \sin kx$ for which $ka \ll 1$. Making the usual assumptions of linear inviscid flow, calculate the perturbation pressure and velocity fields, assuming that the undisturbed flow is parallel to the x direction and has Mach number M. Distinguish the two regimes $M < 1$ and $M > 1$, and for each regime determine the average drag per unit area acting on the surface. If the surface is slightly flexible, in what way is it deformed by the pressure field?

   What happens when $M \to 1$? Obtain the order of magnitude of all of the physical quantities found above when $M \simeq 1$.

8. For small angles of deflection of fluid at an oblique shock when the flow speeds are close to sonic, is there an approximate equation for the shock polar that is consistent with the transonic similarity parameter?

9. Obtain approximate formulae describing Prandtl–Meyer expansion around a corner at Mach number close to 1. Verify that your formulae are consistent with an order-one value of the transonic similarity parameter.

10. Show that the transonic small disturbance equation has a family of similarity solutions $\phi(x, y) = y^a f(x/y^b)$, containing the function f and the constants a and b, provided that f satisfies a certain ordinary differential equation. Can a and b take arbitrary values? If the ordinary differential equation is to be linear, what condition must be satisfied by a and b? Show that a solution in the family is $\phi(x, y) = y^{\frac{2}{5}} f(x/y^{\frac{4}{5}})$, and state the ordinary differential equation then satisfied by f. Does it have solutions expressible in terms of known functions?

11. Starting from Bernoulli's equation and the equation $c^2 \partial u_i / \partial x_i = u_i u_j \partial u_i / \partial x_j$, derive the transonic small disturbance equation for flow past an aerofoil,

and show that the equation may be written in the form

$$(K - (\gamma + 1)\phi_x)\phi_{xx} + \phi_{yy} = 0.$$

Give particular attention to the required scalings, and define carefully all your symbols. Verify that the terms you have neglected are unimportant. Determine the orders of magnitude of the perturbation velocity components and of the lift and wave-drag coefficients.

10

Characteristics for Steady Two-Dimensional Flow

10.1 Governing Equations

In this chapter we apply the theory of characteristics to steady two-dimensional irrotational homentropic flow of an ideal fluid. We use Cartesian coordinates (x, y) and let the corresponding components of the velocity $\mathbf{u}$ be (u, v). The fluid has speed q, density ρ, and sound speed c, and we assume that the fluid is a polytropic gas with ratio of specific heats γ. Thus

$$\mathbf{u} = (u, v), \tag{10.1.1a}$$

$$q^2 = u^2 + v^2, \tag{10.1.1b}$$

$$(\rho u)_x + (\rho v)_y = 0, \tag{10.1.1c}$$

$$u_y - v_x = 0, \tag{10.1.1d}$$

$$\frac{c^2}{\gamma - 1} + \frac{q^2}{2} = \text{const.}, \tag{10.1.1e}$$

$$c^2 \propto \rho^{\gamma - 1}. \tag{10.1.1f}$$

All the results in this chapter are consequences of (10.1.1). We begin by eliminating ρ. Equations (10.1.1e,f) give

$$\frac{\nabla \rho}{\rho} = \frac{1}{\gamma - 1} \frac{\nabla(c^2)}{c^2} = -\frac{\nabla\left(\frac{1}{2}q^2\right)}{c^2}. \tag{10.1.2}$$

Then (10.1.1c) gives

$$\nabla \cdot \mathbf{u} = \frac{-\mathbf{u} \cdot \nabla \rho}{\rho} = \frac{\mathbf{u} \cdot \nabla\left(\frac{1}{2}q^2\right)}{c^2} = \frac{\mathbf{u} \cdot (\mathbf{u} \cdot \nabla \mathbf{u})}{c^2}. \tag{10.1.3}$$

In the last step we have used the irrotationality of the flow. Thus the two basic equations in u and v are

$$(u^2 - c^2)u_x + uv(u_y + v_x) + (v^2 - c^2)v_y = 0, \qquad (10.1.4a)$$

$$u_y - v_x = 0. \qquad (10.1.4b)$$

These form a system of two first-order quasi-linear partial differential equations in u and v. In the next section we determine the characteristics of the system using the methods introduced in Chapter 5.

10.2 The Friedrichs Theory

The simplest method of determining the characteristics of (10.1.4) is the method of linear combinations, as described in the Friedrichs theory of Section 5.2. Let us add to (10.1.4a) an arbitrary multiple λ of (10.1.4b) to obtain

$$\left\{(u^2 - c^2)\frac{\partial}{\partial x} + (uv + \lambda)\frac{\partial}{\partial y}\right\}u + \left\{(uv - \lambda)\frac{\partial}{\partial x} + (v^2 - c^2)\frac{\partial}{\partial y}\right\}v = c.$$

$$(10.2.1)$$

In the (x, y) plane, the differential operator acting on u is a directional derivative in the direction $dy/dx = (uv + \lambda)/(u^2 - c^2)$, and similarly for v in the direction $dy/dx = (v^2 - c^2)/(uv - \lambda)$. The two directions are the same if λ satisfies

$$\frac{uv + \lambda}{u^2 - c^2} = \frac{v^2 - c^2}{uv - \lambda}. \qquad (10.2.2)$$

The common value of dy/dx determined by (10.2.2) will be denoted m, and the angle corresponding to the slope m will be denoted χ, so that $m = \tan \chi$. Thus when λ satisfies (10.2.2) we write

$$m = \frac{dy}{dx} = \tan \chi = \frac{uv + \lambda}{u^2 - c^2} = \frac{v^2 - c^2}{uv - \lambda}. \qquad (10.2.3)$$

The values of m satisfying (10.2.3) are the slopes of the characteristics at a point in the (x, y) plane, as illustrated in Figure 10.2.1.

One method of determining m is to write (10.2.2) as a quadratic equation in λ, and substitute expressions for the roots into (10.2.3). But a simpler method is to eliminate λ from (10.2.3), to obtain directly a quadratic equation for m. Thus we write

$$uv + \lambda = (u^2 - c^2)m, \qquad uv - \lambda = \frac{v^2 - c^2}{m}. \qquad (10.2.4)$$

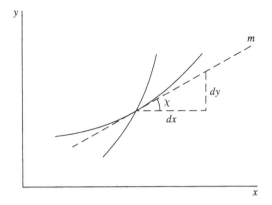

Fig. 10.2.1. Two characteristics (—) through a point. Each characteristic determines a
slope m and angle χ for which $m = dy/dx = \tan \chi$.

The sum of Equations (10.2.4) is

$$2uv = (u^2 - c^2)m + \frac{v^2 - c^2}{m}. \tag{10.2.5}$$

Therefore

$$(u^2 - c^2)m^2 - 2uvm + (v^2 - c^2) = 0. \tag{10.2.6}$$

Since $m = dy/dx$, this equation can be written

$$(v^2 - c^2)\,dx^2 - 2uv\,dx\,dy + (u^2 - c^2)\,dy^2 = 0. \tag{10.2.7}$$

Therefore

$$c^2(dx^2 + dy^2) = (v\,dx - u\,dy)^2. \tag{10.2.8}$$

Define the differential arc length ds by $ds^2 = dx^2 + dy^2$, the flow-direction
angle θ by $\tan \theta = v/u$, and, when the flow is supersonic, the Mach angle μ by
$\sin \mu = c/q$. Then three right-angled triangles are the arc length triangle $(dx,$
$dy, ds)$ with base angle χ, the flow triangle (u, v, q) with base angle θ, and
the Mach triangle $((q^2 - c^2)^{\frac{1}{2}}, c, q)$ with base angle μ. The three triangles are
shown in Figure 10.2.2. In terms of χ and θ, Equation (10.2.8) is

$$c^2\,ds^2 = q^2\,ds^2(\sin \theta \cos \chi - \cos \theta \sin \chi)^2. \tag{10.2.9}$$

Therefore

$$\frac{c}{q} = \pm \sin(\theta - \chi). \tag{10.2.10}$$

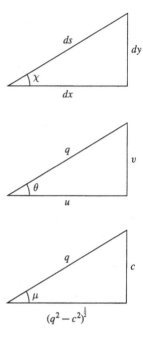

Fig. 10.2.2. Right-angled triangles with base angles obtained from the characteristic direction χ, the stream direction θ, and the Mach angle μ.

Consider a point (x, y) at which the flow is sonic or supersonic (i.e., at which $q \geq c$). Since the Mach angle μ is defined by $\sin\mu = c/q$, it follows that (10.2.10) is $\sin\mu = \pm\sin(\theta - \chi)$, that is,

$$\chi = \theta \pm \mu. \tag{10.2.11}$$

Thus the characteristic directions χ are obtained from the stream direction θ by adding or subtracting the Mach angle μ. These characteristic directions are in agreement with the theorems and physical arguments relating to characteristic surfaces which were presented in Chapter 5. Let the characteristic directions (10.2.11) be denoted χ_+ and χ_-, that is, $\chi_\pm = \theta \pm \mu$ as shown in Figure 10.2.3, and let the corresponding slopes (10.2.3) be denoted m_+ and m_-. Thus

$$m_\pm = \tan\chi_\pm = \tan(\theta \pm \mu). \tag{10.2.12}$$

The expression for the product of the roots of a quadratic in terms of the coefficients, applied to (10.2.6), gives

$$m_+ m_- = \frac{v^2 - c^2}{u^2 - c^2}. \tag{10.2.13}$$

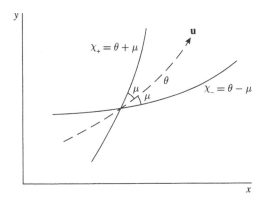

Fig. 10.2.3. Streamline (— — —) in direction θ, and characteristics (—) in directions $\chi_\pm = \theta \pm \mu$, in a flow with velocity **u** and Mach angle μ.

Similarly, some other properties of the slopes $m_\pm$ are obtained more easily from the quadratic equation (10.2.6) than from explicit formuale such as (10.2.12).

10.3 Ordinary Differential Equations on Characteristics

We now reduce the governing equations (10.1.1) to ordinary differential equations on the characteristics, of slopes $m = m_\pm = \tan \chi_\pm = \tan(\theta \pm \mu)$ as we have seen. To obtain an equation for dv/du, we substitute into the linear combination (10.2.1) a value of λ satisfying (10.2.3) and use the differential identity $d/dx = \partial/\partial x + (dy/dx)\,\partial/\partial y$. Thus

$$(u^2 - c^2)\frac{du}{dx} + \left(\frac{v^2 - c^2}{m}\right)\frac{dv}{dx} = 0. \tag{10.3.1}$$

Therefore

$$\frac{dv}{du} = -\left(\frac{u^2 - c^2}{v^2 - c^2}\right)m. \tag{10.3.2}$$

Since c^2 is a known function of u and v, by Bernoulli's equation $c^2/(\gamma - 1) + (u^2 + v^2)/2 = \text{constant}$, so also are the two values $m = m_\pm$, by the relation $m_\pm = \tan(\theta \pm \mu)$, that is,

$$m_\pm = \tan\left(\tan^{-1}\left(\frac{v}{u}\right) \pm \sin^{-1}\left(\frac{c}{q}\right)\right). \tag{10.3.3}$$

Therefore (10.3.2) represents two ordinary differential equations in u and v, one with $m = m_+$ and the other with $m = m_-$. If these equations can be solved, there are functions $F_+(u, v)$ and $F_-(u, v)$ such that the solutions may be written $F_+(u, v) = $ constant and $F_-(u, v) = $ constant. The functions $F_\pm(u, v)$ define new dependent variables, called Riemann variables or Riemann invariants. It is remarkable that (10.3.2), with m given by (10.3.3), can be solved in terms of known functions, so that explicit formulae are available for $F_\pm(u, v)$. We shall obtain these formulae in the next section, but we first discuss the geometry of the characteristics.

Since $m = dy/dx$, let us write (10.3.2) as

$$\frac{dy}{dx} = -\left(\frac{v^2 - c^2}{u^2 - c^2}\right)\frac{dv}{du}. \tag{10.3.4}$$

Substitution into (10.2.8) gives

$$(u^2 - c^2)\,du^2 + 2uv\,du\,dv + (v^2 - c^2)\,dv^2 = 0. \tag{10.3.5}$$

This equation represents the two ordinary differential equations in u and v that would be obtained by substituting $m = m_+$ and $m = m_-$ into (10.3.2).

Equation (10.3.2) may also be written, by (10.2.13), as

$$\frac{dv}{du} = -\frac{m}{m_+ m_-}. \tag{10.3.6}$$

Here m is either member of the pair $m_\pm$, and the other member may be denoted m'. Thus (m, m') is either (m_+, m_-) or (m_-, m_+). The corresponding characteristic directions will be denoted χ and χ', so that $m = \tan \chi$ and $m' = \tan \chi'$. Equation (10.3.6) is therefore $dv/du = -1/m'$; thus

$$\left(\frac{dv}{du}\right)\left(\frac{dy}{dx}\right)' = -1. \tag{10.3.7}$$

This is equivalent to $(dv/du) \tan \chi' = -1$. Since two lines are perpendicular when the product of their slopes is -1, Equation (10.3.7) asserts that each characteristic is perpendicular to the line with slope equal to the value of dv/du on the other characteristic. Recall that (dx, dy) and (du, dv) are changes in (x, y) and (u, v) on a characteristic through (x, y), not on a streamline. The slope of the streamline through (x, y) is $\tan \theta$ (i.e., v/u). Thus (10.3.7) makes no assertion about neighbouring points on a streamline nor that a streamline is perpendicular to any specified direction. The details of (10.3.7) for each characteristic are illustrated in Figure 10.3.1.

The formulae of this section take a simple form when the velocity is expressed in polar coordinates. We put $(u, v) = (q \cos\theta, q \sin\theta)$, so that $(du, dv) =$

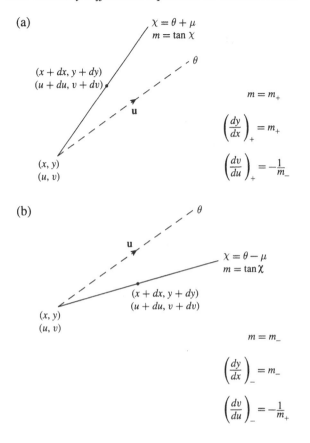

Fig. 10.3.1. Values of dy/dx and dv/du on characteristics: (a) $m = m_+$; (b) $m = m_-$. The basic relations are $(dy/dx)_+ (dv/du)_- = -1$ and $(dy/dx)_- (dv/du)_+ = -1$.

$(dq \cos \theta - q \sin \theta \, d\theta, \, dq \sin \theta + q \cos \theta \, d\theta)$. Then (10.3.7) in the form $\tan \chi' = -du/dv$ is

$$\tan \chi' = -\frac{dq \cos \theta - q \sin \theta \, d\theta}{dq \sin \theta + q \cos \theta \, d\theta}. \qquad (10.3.8)$$

We may put $\chi = \theta \pm \mu$ and $\chi' = \theta \mp \mu$, so that

$$\tan(\theta \mp \mu) = \frac{\tan \theta - q^{-1} dq/d\theta}{1 + (\tan \theta) q^{-1} dq/d\theta}. \qquad (10.3.9)$$

The identity $\tan(\theta \mp \mu) = (\tan \theta \mp \tan \mu)/(1 \pm \tan \theta \tan \mu)$ shows that (10.3.9) is equivalent to $\mp \tan \mu = -q^{-1} dq/d\theta$, that is,

$$\pm d\theta = (\cot \mu) \frac{dq}{q}. \qquad (10.3.10)$$

The quantity $q^{-1}dq$ may be expressed in terms of μ and $d\mu$, for by Bernoulli's equation the flow takes place at a constant value of $c^2/(\gamma - 1) + \frac{1}{2}q^2$, so that, since $\sin \mu = c/q$, we have

$$\left(\sin^2\mu + \tfrac{1}{2}(\gamma - 1)\right)q^2 = \text{const.} \tag{10.3.11}$$

The logarithmic differential of this equation is

$$\frac{\sin \mu \cos \mu \, d\mu}{\sin^2 \mu + \tfrac{1}{2}(\gamma - 1)} + \frac{dq}{q} = 0. \tag{10.3.12}$$

Elimination of $q^{-1}dq$ from (10.3.10) and (10.3.12) gives

$$\mp d\theta = \frac{\cos^2 \mu \, d\mu}{\sin^2 \mu + \tfrac{1}{2}(\gamma - 1)}. \tag{10.3.13}$$

Thus we have obtained an ordinary differential equation relating two important angles in the flow, namely the stream angle θ and the Mach angle μ. We solve the equation in the next section.

10.4 Riemann Invariants

The Riemann invariants are the integrals of (10.3.13). Substituting $t = \tan \mu$, we obtain

$$
\begin{aligned}
\mp \int d\theta &= \int \left(\frac{\tfrac{1}{2}(\gamma + 1)}{\sin^2 \mu + \tfrac{1}{2}(\gamma - 1)} - 1 \right) d\mu \\
&= \int \frac{dt}{t^2 + (\gamma - 1)/(\gamma + 1)} - \mu \\
&= \left(\frac{\gamma + 1}{\gamma - 1} \right)^{\frac{1}{2}} \tan^{-1} \left\{ \left(\frac{\gamma + 1}{\gamma - 1} \right)^{\frac{1}{2}} \tan \mu \right\} - \mu.
\end{aligned}
\tag{10.4.1}
$$

This expression in μ defines the Prandtl–Meyer function $P(\mu)$. Thus

$$P(\mu) = \left(\frac{\gamma + 1}{\gamma - 1} \right)^{\frac{1}{2}} \tan^{-1} \left\{ \left(\frac{\gamma + 1}{\gamma - 1} \right)^{\frac{1}{2}} \tan \mu \right\} - \mu. \tag{10.4.2}$$

We defined in (8.2.18) the Prandtl–Meyer function $\nu(M)$. Since $\tan \mu = 1/(M^2 - 1)^{\frac{1}{2}}$, we obtain from the trigonometric identity $\tan^{-1}(1/x) = \frac{1}{2}\pi - \tan^{-1}(x)$ the relation $P(\mu) = (\{(\gamma + 1)/(\gamma - 1)\}^{\frac{1}{2}} - 1)\pi/2 - \nu(M)$. Thus $P(\mu)$ and $\nu(M)$, with $\sin \mu = 1/M$, though simply related are not quite the same quantity expressed in different variables. The context makes clear whether the term Prandtl–Meyer function refers to $P(\mu)$ or $\nu(M)$. In this chapter, the more useful function is $P(\mu)$.

Since (10.4.1) applies on the characteristics $\chi = \theta \pm \mu$, we have

$$\theta \pm P(\mu) = \text{const.} \quad \text{on} \quad \frac{dy}{dx} = \tan(\theta \pm \mu). \tag{10.4.3}$$

Using (10.3.3), we may write this equation in terms of u and v to arrive at known functions $F_\pm(u, v)$ and $f_\pm(u, v)$ for which

$$F_\pm(u, v) = \text{const.} \quad \text{on} \quad \frac{dy}{dx} = f_\pm(u, v). \tag{10.4.4}$$

Thus we have shown that $\theta \pm P(\mu)$, or equivalently $F_\pm(u, v)$, are the two Riemann invariants for the system of governing equations (10.1.1) describing steady two-dimensional irrotational homentropic flow of an ideal polytropic gas with ratio of specific heats γ. The Riemann invariants are defined only where the Mach angle μ is defined, that is, in regions where the flow is sonic or supersonic. In such regions, the Riemann invariants can be made the basis of an effective computational procedure for determining the flow.

10.5 Flow round a Smooth Bend

We now apply the theory of characteristics to supersonic flow round a smooth convex bend. We use a coordinate system with the x axis horizontal and the y axis vertical, and we consider flow above a wall that upstream is straight and parallel to the x axis, and downstream curves towards decreasing y. In terms of a parameter ξ and functions $\hat{y}(\xi)$ and $\hat{\theta}(\xi)$, the wall is the curve $(x, y) = (\xi, \hat{y}(\xi))$ and the angle between the wall and the x axis is $\hat{\theta}(\xi)$. At an arbitrary position in the fluid, the flow is at an angle θ to the x axis, so that on the wall $\theta = \hat{\theta}(\xi)$. Assume that upstream the fluid is uniform and in rectilinear motion at constant speed, with pressure p_0, speed of sound c_0, horizontal component of velocity u_0, vertical component of velocity zero, flow speed q_0, flow angle θ_0, Mach number M_0, and Mach angle μ_0. Then the upstream conditions are

$$(u, v) = (u_0, 0),$$
$$(q, \theta) = (q_0, \theta_0) = (u_0, 0), \tag{10.5.1}$$
$$(M, \mu) = (M_0, \mu_0) = \left(\frac{u_0}{c_0}, \sin^{-1}\left(\frac{c_0}{u_0} \right) \right).$$

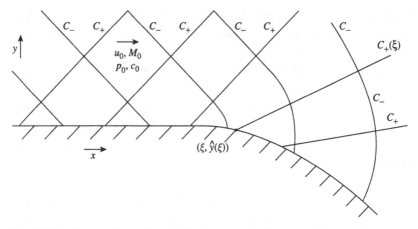

Fig. 10.5.1. Characteristics C_+ and C_- in supersonic flow round the smooth convex bend $(x, y) = (\xi, \hat{y}(\xi))$. The upstream fluid has velocity $(u_0, 0)$, Mach number M_0, pressure p_0, and speed of sound c_0.

The upstream characteristic angles and slopes are

$$\chi_\pm = \theta_0 \pm \mu_0 = \pm\mu_0,$$
$$\left(\frac{dy}{dx}\right)_\pm = m_{0\pm} = \pm\tan\mu_0.$$
(10.5.2)

Several of these quantities are shown in Figure 10.5.1.

In (10.4.3), let the characteristics for which $\theta + P(\mu) = $ constant and $dy/dx = \tan(\theta+\mu)$ be denoted C_+, and let those for which $\theta - P(\mu) = $ constant and $dy/dx = \tan(\theta - \mu)$ be denoted C_-. In most problems the constant values of $\theta + P(\mu)$, and of $\theta - P(\mu)$, differ from one characteristic to another. But Figure 10.5.1 shows that, in our problem, all the C_- characteristics originate in the upstream region of uniform conditions. Therefore $\theta - P(\mu)$ takes the constant value $\theta_0 - P(\mu_0)$ (i.e., $-P(\mu_0)$) in the whole flow. Thus everywhere we have

$$\theta - P(\mu) = -P(\mu_0).$$
(10.5.3)

A flow in which each Riemann invariant except one is constant throughout the flow region is said to be a simple wave. We have thus shown that the flow of an initially uniform stream round a smooth convex bend is a simple wave.

The C_+ characteristic that meets the wall at $x = \xi$ will be denoted $C_+(\xi)$. Its equation may be written in terms of a function $f_+(\xi)$, as yet unknown, as

$$\theta + P(\mu) = f_+(\xi).$$
(10.5.4)

Equations (10.5.3) and (10.5.4) imply that θ and μ are functions of ξ, say $\hat{\theta}(\xi)$ and $\hat{\mu}(\xi)$, where $\hat{\theta}(\xi)$ is the function introduced at the beginning of this section. Thus

$$(\theta, \mu) = (\hat{\theta}(\xi), \hat{\mu}(\xi)). \tag{10.5.5}$$

Define $\hat{m}(\xi)$ by

$$\hat{m}(\xi) = \tan(\hat{\theta}(\xi) + \hat{\mu}(\xi)). \tag{10.5.6}$$

Then by (10.4.3), the differential equation for the position of $C_+(\xi)$ is

$$\frac{dy}{dx} = \tan(\hat{\theta}(\xi) + \hat{\mu}(\xi)) = \hat{m}(\xi). \tag{10.5.7}$$

But ξ is constant on the characteristic $C_+(\xi)$. Therefore (10.5.7) shows that the C_+ characteristics are straight lines. Since $C_+(\xi)$ meets the wall at $(x, y) = (\xi, \hat{y}(\xi))$, its equation is

$$y = \hat{y}(\xi) + \hat{m}(\xi)(x - \xi). \tag{10.5.8}$$

We have still to determine the functions $\hat{\mu}(\xi)$ and $\hat{m}(\xi)$. The starting point is $\hat{\theta}(\xi)$, given from the equation of the wall by $\tan \hat{\theta}(\xi) = \hat{y}'(\xi)$. Then (10.5.3) gives $\hat{\mu}(\xi)$, and (10.5.6) gives $\hat{m}(\xi)$. In principle, (10.5.8) may then be inverted to give ξ as a function of x and y, so that the functions $\theta = \hat{\theta}(\xi)$ and $\mu = \hat{\mu}(\xi)$ determine θ and μ as functions of x and y. All quantities can now be found as functions of position. Thus the flow speed q is obtained from Bernoulli's equation $(\sin^2 \mu + \frac{1}{2}(\gamma - 1))q^2 = $ constant; the velocity components from $(u, v) = (q \cos \theta, q \sin \theta)$; the speed of sound from $c = q \sin \mu$; and thermodynamic quantities from their dependence on c, as the flow is homentropic. We thus obtain the complete solution of the full nonlinear problem of supersonic flow round a smooth convex bend. Such a complete solution is often available for a flow that is a simple wave.

The theory of characteristics makes precise the idea, introduced in Section 8.3, of the flow as a sequence of Prandtl–Meyer expansions of infinitesimal strength. The two approaches are equivalent, as is evident geometrically from a comparison of Figures 8.3.1 and 10.5.1, and algebraically from the common occurrence of the Prandtl–Meyer function. Thus the Prandtl–Meyer expansion, though introduced in relation to the special problem of flow round a sharp corner, provides for more general high speed flows a building block of some versatility.

The theory of characteristics may also be applied to flow round a concave bend. But now the C_+ characteristics converge and cross each other at a certain distance from the wall, and the flow contains a shock. For example, if the right half of the wall in Figure 10.5.1 curves upwards, then the C_+ characteristics are still straight lines, but they now typically have a cusped caustic, and a shock forms near the region of the cusp. We discussed this type of flow in Section 8.3 and sketched it in Figure 8.3.2. The region in which the flow is a simple wave is bounded downstream by the shock and by the C_- characteristic from the upstream end of the shock.

Many other types of supersonic flow, for example past an aerofoil or inside a jet engine, can be analysed by the theory of characteristics. The position of a shock may be hard to calculate, not least because its strength and curvature are partly affected by any incoming characteristics that reflect off it. Such complicated flows usually require numerical investigation. The theory of characteristics forms the basis of many numerical schemes for supersonic flow, which often incorporate finite-difference approximations to the differential equations presented in this chapter.

10.6 Bibliographic Notes

The classic early paper on characteristics, for unsteady one-dimensional flow, is Riemann (1860). Reference articles in Table 13.3.1 that include the theory of characteristics in two independent variables are Meyer (1953), Ferri (1955a), Friedrichs (1955), Kantrowitz (1958), and Meyer (1960); texts and monographs in Table 13.4.1 that cover this material are Courant & Friedrichs (1948) and Whitham (1974). See also Courant & Hilbert (1962). The theory of characteristics in two independent variables forms part of many of the recent research papers listed in Tables 13.6.1a,c,d, especially those on shocks, jets, aerofoils, and ducts.

10.7 Further Results and Exercises

1. Write down the equations of motion for steady inviscid flow of an isentropic fluid with axial symmetry, using cylindrical polar coordinates with radial coordinate r and axial coordinate z. Show that the characteristics are the streamlines and the Mach lines.

   Assume now that the fluid is a polytropic gas with ratio of specific heats γ. Let the gas have pressure p, Mach angle μ, speed q, and axial velocity component w, and let the angle between the flow and the axis be θ. Show that, on the Mach lines,

   $$\sin \mu \cos \mu \frac{dp}{\gamma p} + \frac{\sin \mu \, \sin \theta}{\sin(\theta \pm \mu)} \frac{dr}{r} \pm d\theta = 0.$$

   Why does this relation fail on the axis? Show that the corresponding relation is then

$$\frac{dq}{r} \mp q \tan \mu \frac{d\theta}{dr} - \frac{1}{2} \frac{\cos \mu}{\sin(\theta \pm \mu)} \frac{\partial w}{\partial z} = 0.$$

2. Express in characteristic form the equations

$$(u^2 + v^2)(u_x - v_y) - (u^2 - v^2)(u_y + v_x) = 0,$$

$$(u^2 + v^2)(u_y - v_x) + (u^2 - v^2)(u_x + v_y) = 0.$$

Find the region in the quadrant $x \geq 0$, $y \geq 0$ in which the values of u and v are determined by the boundary condition $(u, v) = (1, 1)$ on the line $y = 0, x \geq 0$. Show that the values of u and v are determined at all points of the quadrant $x \geq 0$, $y \geq 0$ if, in addition, u satisfies the boundary condition $u = 2/(y + 2)$ on the line $x = 0$, $y \geq 0$. Show that the family of characteristic lines is then given in terms of a positive parameter y_0 by the equation

$$y = y_0 + (y_0 + 1)^2 x.$$

Find the values of u and v at the point $(x, y) = (1, 5)$.

3. Air flows supersonically round a convex wall consisting of two straight sections connected by a 90° circular arc. Upstream and downstream the velocity is uniform and parallel to the wall. Describe the flow as completely as you can, giving particular attention to the dependence of velocity, pressure, and sound speed on position, assuming that boundary-layer effects may be ignored. What would happen if the two straight sections of the wall were connected by a 180° circular arc?

4. The governing equations for steady two-dimensional irrotational homentropic flow of an ideal polytropic gas may be written, in standard notation, as

$$c^2(u_x + v_y) = u^2 u_x + uv(u_y + v_x) + v^2 v_y,$$

$$u_y - v_x = 0.$$

Let the ratio of specific heats of the gas be γ, and let the Mach angle of the flow, defined in supersonic regions and in general a function of position, be μ. Show that application of the two-dimensional theory of characteristics to the governing equations leads in a natural way to the function $P(\mu)$

defined by

$$P(\mu) = \left(\frac{\gamma+1}{\gamma-1}\right)^{\frac{1}{2}} \tan^{-1}\left\{\left(\frac{\gamma+1}{\gamma-1}\right)^{\frac{1}{2}} \tan\mu\right\} - \mu.$$

Give a brief account of how the function $P(\mu)$ determines the flow in a supersonic region.

5. Describe in detail how the method of characteristics may be used to determine the two-dimensional supersonic flow of air round a smooth convex bend of prescribed shape. In your working, denote the flow angle by θ, the Mach angle by μ, and the Prandtl–Meyer function by $P(\mu)$. You do not need to derive the functional form of $P(\mu)$, and you may assume that the Riemann invariants for the flow are $\theta \pm P(\mu)$. Show how the velocity components and all thermodynamic quantities could in principle be calculated as a function of position.

What happens if the bend is concave? How would you then find the region in which the flow is determined in a simple way by the conditions upstream? Draw a sketch of the whole flow field.

6. Write an account of the theory of simple waves for steady two-dimensional supersonic flow of a polytropic gas.

7. Describe the general differences between steady subsonic and supersonic flow of a fluid, as revealed by an examination of (a) the variation of cross-sectional area along a stream tube; (b) the equation for the potential in irrotational flow; and (c) any other considerations.

8. State the equations describing unsteady one-dimensional homentropic flow of an ideal polytropic gas. Show that, in the usual notation, the Riemann invariants are $c/(\gamma - 1) \pm u/2$ and they are constant on the characteristics $dx/dt = \pm c + u$.

The gas is initially at rest in a uniform state, with speed of sound c_0, in a semi-infinite cylinder aligned with the positive x axis. At time $t = 0$, a piston at $x = 0$ starts to move with constant acceleration f in the negative x direction (i.e., away from the gas). Show that the piston separates from the gas after a time $t = (2/(\gamma - 1))c_0/f$.

What would be the motion of the gas if from $t = 0$ the piston moved with constant acceleration f in the positive x direction (i.e., towards the gas)? Calculate the velocity field in this case, and hence or otherwise show that a discontinuity forms in the gas after a time $t = (2/(\gamma - 1))c_0/f$. Where is the piston when the discontinuity first appears?

9. In unsteady one-dimensional homentropic flow of an ideal polytropic gas, let coordinates based on the characteristics be denoted α and β, so that one family of characteristics is given by constant α, and the other by constant β.

Show that the equations of motion, in the usual variables x, t, u, c, and γ, may be written

$$\frac{\partial x}{\partial \alpha} = (c + u)\frac{\partial t}{\partial \alpha}, \qquad \frac{\partial}{\partial \alpha}\left(\frac{c}{\gamma - 1} + \frac{u}{2}\right) = 0,$$

$$\frac{\partial x}{\partial \beta} = (-c + u)\frac{\partial t}{\partial \beta}, \qquad \frac{\partial}{\partial \beta}\left(\frac{c}{\gamma - 1} - \frac{u}{2}\right) = 0.$$

Let the initial state of the gas at $t = 0$, in dimensionless variables and in terms of a small parameter ϵ, be

$$\alpha = \beta = x, \qquad u = 0, \qquad c = 1 + \epsilon \cos x.$$

Show that a discontinuity first appears in the gas at a time of order ϵ^{-1} and at a position for which, with error of order 1, the characteristic coordinates α and β satisfy

$$(\alpha - \beta)(\sin \alpha - \sin \beta) = -\frac{4}{\epsilon}\left(\frac{\gamma - 1}{\gamma + 1}\right).$$

11

Shock Reflections and Intersections

11.1 Common Shock Patterns

The two basic shock patterns found at a reflection of a shock off a wall are the regular reflection, consisting of an incident and a reflected shock, and the Mach reflection, containing also a Mach shock and a vortex sheet. A section of a Mach shock is a Mach stem, and a section of a vortex sheet is a slip line. In sections, the incident and reflected shocks, the Mach stem, and the slip line meet at the triple point. The two patterns occur in steady flow (e.g., when a shock in a wind tunnel reflects off a wall), in pseudo-steady flow (i.e., self-similar flow, e.g., when a shock is diffracted by a wedge or by a corner in a piecewise planar wall), and in unsteady flow (e.g., when a shock is diffracted by a convex or concave cylinder or sphere, or by a corner and curved section in a wall). The patterns are illustrated in Figure 11.1.1.

A Mach stem is usually curved; it meets the wall at right angles, to keep the flow tangential to the wall. For the same reason, any single shock that meets a wall does so at right angles. At an intersection of shocks in a region remote from a wall, the corresponding shock patterns are the regular intersection and the Mach intersection, shown in Figure 11.1.2. The Mach shock in an axisymmetric pattern is called a Mach disc. To a reflection there corresponds a symmetric intersection, but without boundary layer effects.

Variants of the basic shock patterns are shown in Figure 11.1.3. In (a), the Mach stem is a normal shock perpendicular to the wall, so that the slip line is parallel to the wall. The pattern allows a Mach stem of any length, including zero, and so includes regular reflection as a special case. The occurrence of this pattern, in which the flow is said to satisfy the mechanical equilibrium condition, thus provides a possible transition criterion, known as the von Neumann criterion, for smooth transition between Mach and regular reflection. Shock pattern (a) is called a stationary Mach reflection. In (b), the Mach stem is on the upstream side of the triple point, and so it deflects the nearby flow away

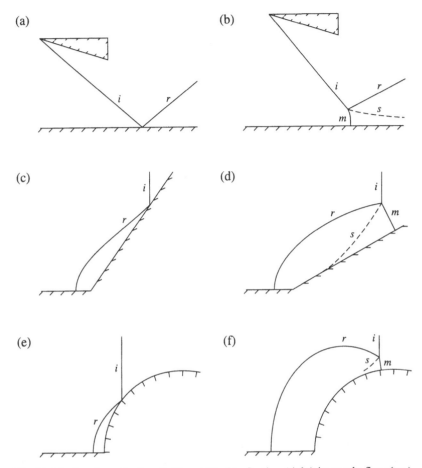

Fig. 11.1.1. Regular reflection (left) and Mach reflection (right) in steady flow (top), pseudo-steady flow (center), and unsteady flow (bottom). i, incident shock; r, reflected shock; m, Mach stem; s, slip line. In the top row the shocks are stationary and the flow is from left to right; in the lower two rows the incident shock is moving from left to right into stationary (i.e., quiescent) fluid. Unsteady patterns similar to (e), (f) would be produced if the curved part of the boundary were concave.

from the wall, in accordance with the flow-deflection property of an oblique shock illustrated in Figure 6.4.1. Therefore the slip line diverges from the wall. Note that in our diagrams of instantaneously steady flow, the flow upstream of the Mach stem and incident shock is parallel to the wall. Shock pattern (b) is called an inverted, or inverse, Mach reflection. In (c) the reflected shock slopes upstream, and in (d) the slip line has been replaced by a narrow fan-shaped mixing region. In (e), there is no reflected shock, but the mainly supersonic flow downstream of the incident shock contains a subsonic patch next to the wall

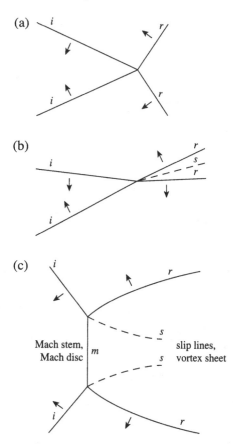

Fig. 11.1.2. (a) Symmetric regular shock intersection. (b) Asymmetric regular shock intersection, with slip line. (c) Symmetric Mach intersection. In two dimensions, a Mach stem joins two triple points, from each of which emerges a slip line. In three dimensions, a Mach disc is bounded by a Mach rim from which emerges a tubular vortex sheet.

near the foot of the shock. We include (e) to point out that not every incident shock produces a reflected shock. In (f), the reflected shock has a reversal of curvature, which generates a compression wave. The pattern is a transitional Mach reflection. In (g), the smooth compression wave has shrunk to zero width (i.e., has become a shock), and the reversal of curvature has become a kink that is the triple point of a second Mach reflection. The pattern is a double Mach reflection. In (h), the triple point of a Mach reflection has been replaced by a region containing a single smoothly curved shock next to a smooth compression wave that steepens further out to become the reflected shock. This pattern, which is difficult to distinguish experimentally from a Mach reflection when the compression wave is thin, is a von Neumann reflection. In (i), the Mach stem

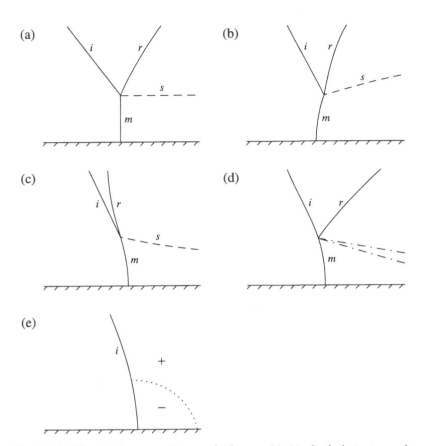

Fig. 11.1.3. Other shock patterns. Frames of reference: (a)–(e), shocks instantaneously stationary and flow from left to right; (f)–(i), incident shock moving from left to right into stationary fluid. (a) Stationary Mach reflection, satisfying the von Neumann criterion (i.e., the mechanical-equilibrium criterion). Mach stem straight and perpendicular to wall; slip line straight and parallel to wall. (b) Inverse Mach reflection. Foot of Mach stem on upstream side of triple point; slip line diverging from wall. (c) Reflected shock sloping upstream. Possible in theory, but not observed. (d) Angular (i.e., fan-shaped) mixing region (with boundaries –·–·), instead of slip line. (e) No reflected shock. The sonic line (···) bounds a subsonic patch (−) surrounded by a supersonic region (+). (f) Transitional Mach reflection. Reversal of curvature in reflected shock, and a fan-shaped smooth compression wave (c). (g) Double Mach reflection. Kink in reflected shock forms triple point of second Mach reflection (i', r', m', s'). (h) Von Neumann reflection. Reflected shock produced by steepening of smooth compression wave (c); incident shock and Mach stem form one continuous curve, without jump in slope. (i) Guderley supersonic patch (+) surrounded by subsonic region (−). Boundary is sonic line (···).

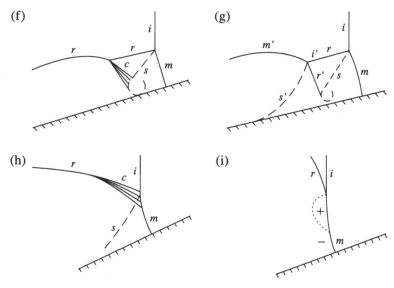

Fig. 11.1.3. (*Continued*)

has supersonic flow behind the segment that is most oblique (i.e., furthest from the wall), in accordance with the theory of oblique shocks given in Section 6.4. The pattern is said to contain a Guderley supersonic patch.

Some of these patterns can be classified further. In the double Mach reflection (g), for example, the straight shock ri' joining the two triple points may point towards the vertex of the wedge that generated the shock pattern or it may point above or below this vertex. Other patterns we have not shown include variants of (f) and (g) containing a regular reflection.

A great deal is known about the conditions in which the different patterns occur and about the criteria for transitions between them, including hysteresis effects. This knowledge is not easily summarised, because of the complicated dependence of the patterns on many parameters. These include incident shock strength and angle, equivalent to incident Mach number and angle; wedge angle; whether the flow is steady, pseudo-steady, or unsteady; ratio of specific heats; real gas effects, for example viscosity, thermal conductivity, and atomic and molecular vibration, dissociation, electronic excitation, and ionization; three-dimensional effects, often aspect-ratio or cross-flow effects; surface roughness, porosity, and perforations; fluctuations in the upstream flow; and downstream conditions. The last of these is most important, as atypical or artificial patterns can be created by downstream obstructions. More generally, it is important to know whether a given configuration is determined locally, or whether remote boundaries have an influence. In a Mach reflection, of any type, the length of the Mach stem is left indeterminate by local conditions. Thus a Mach reflection can occur only if there exists a pathway, at no point requiring upstream propagation against a supersonic flow, by which information about a length scale in the boundary conditions can be communicated to the Mach stem.

With the above reservations in mind, we may nevertheless say that when the angle between the incident shock and the wall is not too great, the reflection is regular; but if the angle is increased then a transition occurs to Mach reflection. In steady flow with a strong incident shock, the transition usually occurs in accordance with the von Neumann criterion (i.e., at the mechanical equilibrium condition illustrated in Figure 11.1.3a); if the flow is three dimensional, for example because of an aspect ratio effect, then the transition in the other direction, from Mach reflection to regular reflection, often occurs when the shock configuration satisfies the detachment condition, as defined in Section 11.5. Thus the transition displays hysteresis. If the flow is two dimensional or closely so, hysteresis occurs only if care is taken to minimise fluctuations in the upstream flow; otherwise the transition in each direction occurs in accordance with the von Neumann criterion. In pseudo-steady flow with a strong incident shock, the transition usually occurs when the flow behind the reflected shock passes through the sonic velocity, relative to the intersection point of the shocks – that is, when the downstream flow, which was supersonic, becomes subsonic, and so allows conditions downstream to influence the reflection. If the incident shock is not strong, the transition criterion is different. For example, a weak incident shock can undergo a transition to the von Neumann reflection illustrated in Figure 11.1.3h. The most recent studies, with accurate experimental or numerical control of cross-flow and upstream fluctuations, provide increasing evidence for hysteresis effects.

11.2 The von Neumann–Henderson Theory

A method of analysing shock reflections and intersections is to assume at the outset a definite type of configuration (e.g., a regular reflection) and then determine, from the oblique shock relations, the number of possible configurations of that type for given values of certain parameters (e.g., upstream Mach number M_1 and incident shock angle ϕ_1). The process may be repeated for other types of configuration, for example a Mach reflection, and the results combined. Deductions can then be made about the number and type of possible configurations for all parameter values in some range of interest, for example all (M_1, ϕ_1) with $M_1 \geq 1$ and $0 \leq \phi_1 \leq \frac{1}{2}\pi$. The method, initiated by von Neumann and extended by Henderson, leads to the von Neumann–Henderson theory of shock reflection and intersection. The method cannot give complete results because "possible" does not imply "observed in practice," merely "consistent with the oblique shock relations," and some configurations, for instance regular

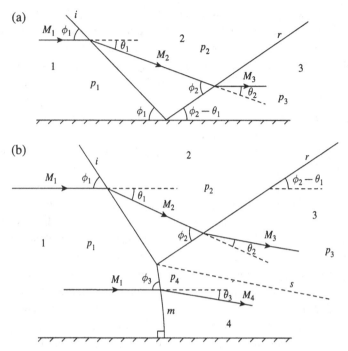

Fig. 11.2.1. Mach numbers, angles, and pressures for (a) regular reflection and (b) Mach reflection. The angles θ_2 and θ_3, which may take either sign, are positive when as shown. The angle of incidence is ϕ_1, and the angle of reflection is $\phi_2 - \theta_1$.

reflection with the stronger of the two possible reflected shocks, are not usually observed. It is nevertheless useful to determine all the possible configurations, because the subject can be viewed as part of bifurcation theory, and theorems are available about the way in which different branches of solutions can join up.

In Sections 11.3–11.14 we apply the method to give an account of the possible configurations for regular and Mach reflection. As indicated in Section 11.15, the method could be applied to some of the more complicated reflections, but the algebra would be rather burdensome. The notation for a regular reflection is shown in Figure 11.2.1a and for a Mach reflection in Figure 11.2.1b. For each we use a frame of reference in which the shocks are at rest and the flow is from left to right. The flow contains an upstream region 1, a central region 2, and a downstream region 3, behind the reflected shock. A Mach reflection also has a downstream region 4, behind the Mach stem, separated from region 3 by a slip line.

Our technique will be to express all quantities in terms of M_1 and ϕ_1 and determine lines in the (M_1, ϕ_1) plane that separate configurations with different properties. Some of the lines are given by a formula for ϕ_1 as a function of M_1; others are given by a contour of a function of M_1 and ϕ_1. In each case,

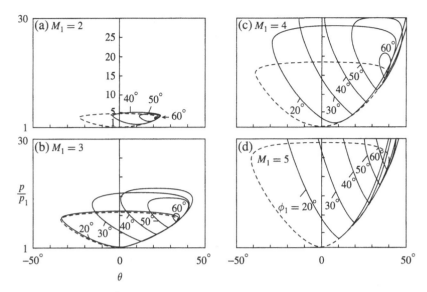

Fig. 11.2.2. Pressure-deflection polars in the $(\theta, p/p_1)$-plane for a polytropic gas with $\gamma = 1.4$. $---$, one-parameter family of incident-shock polars, labelled by upstream Mach number M_1; ———, two-parameter family of reflected-shock polars, labelled by M_1 and incident-shock angle ϕ_1, marked on the polars.

the lines may be plotted using standard graphics software. We always assume that $M_1 \geq 1$ and $0 \leq \phi_1 \leq \frac{1}{2}\pi$, to ensure that the configuration contains an incident shock. Our main analytical tool is the set of equations for the pressure-deflection shock polars in Figure 11.2.2. The regular reflection is analysed in Sections 11.3–11.7 and the Mach reflection in Sections 11.8–11.12. We have summarised the results in Figure 11.2.3, with its rather long caption. Coordinates (M_1, ϕ_1) of particular points in Figure 11.2.3a,b are given in Table 11.2.1. We comment further on regular and Mach reflection in Sections 11.13 and 11.14 and on more general configurations in Section 11.15.

11.3 Regular Reflection

We begin with the regular reflection in Figure 11.2.1a. At fixed $M_1 \geq 1$, Equations (6.4.15) and (6.4.5b) show that the pressure-deflection polar for the incident shock is given parametrically in terms of ϕ_1 by

$$\tan\theta_1 = \frac{\left(M_1^2 \sin^2\phi_1 - 1\right)\cot\phi_1}{1 + M_1^2\left(\frac{1}{2}(\gamma + 1) - \sin^2\phi_1\right)}, \tag{11.3.1a}$$

$$\frac{p_2}{p_1} = 1 + \frac{2\gamma}{\gamma + 1}\left(M_1^2 \sin^2\phi_1 - 1\right). \tag{11.3.1b}$$

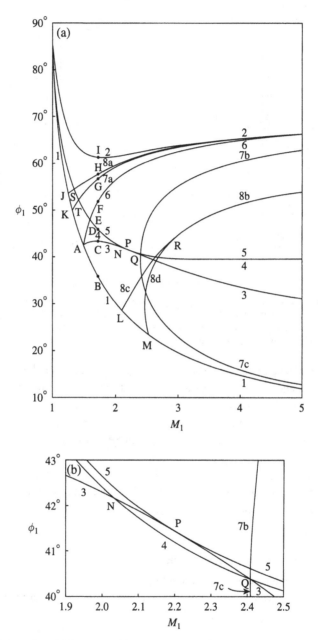

Fig. 11.2.3. (a) Special lines 1–8 in the (M_1, ϕ_1) plane for regular and Mach reflection in a polytropic gas with $\gamma = 1.4$; (b) tangency of lines 3 and 5 at the von Neumann point P, and common intersection of 3, 4, 7b and 7c at the second Henderson point Q. In (a), lines 4 and 5 are scarcely distinguishable.

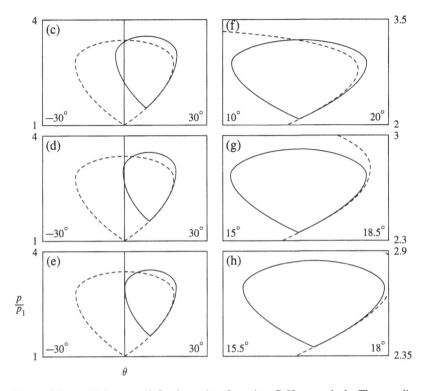

Fig. 11.2.3. (c)–(h) Pressure-deflection polars for points C–H respectively. The coordinates of the marked points are given in Table 11.2.1. (1) Mach-wave incident shock; $\phi_1 = \sin^{-1}(1/M_1), \theta_1 = 0$. Same as curved part of contour $\theta = 0$ in Figure 6.4.4b. Reflected and incident polars are coincident. See Section 11.6. (2) Exactly sonic flow behind incident shock; $M_2 = 1$. Same as curve $M_2 = 1$ in Figure 6.4.4b. Reflected polar has shrunk to a point at right sonic point of incident polar. See Section 11.4. (3) von Neumann line; $\phi_3 = \frac{1}{2}\pi, \theta_3 = 0$. Regular and Mach reflection simultaneously. Above the line: $\theta_3 > 0$, direct Mach reflection; below: $\theta_3 < 0$, inverse. To the left of the von Neumann point P, polars as in (c); to the right, as in Figure 11.2.2d, but with $\phi_1 \simeq 31°$; at P, as in (e), but with the vertical-tangent point raised to the dashed line. See Section 11.10. (4) Sonic line for regular reflection; $M_3 = 1$. Above the line, in narrow strip below 5: $M_3 < 1$; below the line: $M_3 > 1$. Left sonic point of reflected polar is on pressure axis. To the left of N, polars as in (d); to the right, as obtainable from Figure 11.2.2. See Section 11.7. (5) Detachment line for regular reflection; $\theta_2 = \theta_{2,\text{max}}$. Reflected polar tangential to pressure axis, as in (e). See Section 11.5. (6) Normal reflected-shock line for Mach reflection; $\phi_2 = \frac{1}{2}\pi, \theta_2 = 0$. To the right of A: $\phi_3 < \frac{1}{2}\pi$; at A: $\phi_3 = \frac{1}{2}\pi$. Top of reflected polar is on incident polar, as in (f). See Section 11.12. (7) Henderson lines, that is, sonic lines for Mach reflection; $M_3 = 1$. Left or right sonic point of reflected polar is on incident polar. First Henderson line, 7a: polars as in (g), $\phi_2 > \frac{1}{2}\pi, \phi_3 \le \frac{1}{2}\pi$; second, 7b,c: polars obtainable from Figure 11.2.2; 7b: <, ≤; 7c: <, ≥; first Henderson point, T; second, Q; first Henderson region, between 7a and 8a; second, right of 7b,c. In Henderson regions: $M_3 > 1$; between: $M_3 < 1$. See Section 11.11. (8) Change in number n of Mach-reflection configurations. Above 8a: von Neumann region, $n = 0$; between 8a and 8b,c: $n = 1$; below 8b,d: $n = 2$; between 8c and 8d: $n = 3$. Reflected polar tangential to incident polar; point of tangency is vertex of reflected polar, except for 8c. On 8a, polars as in (h); on 8b–d, as obtainable from high resolution of Figure 11.2.2. See Section 11.9.

Table 11.2.1. *Coordinates*
(M_1, ϕ_1) *of points in Figures*
11.2.3a,b. P, *von Neumann point;*
Q, *second Henderson point;*
T, *first Henderson point.*

	M_1	ϕ_1 (degrees)
A	1.48	42.40
B	1.75	34.85
C	1.75	42.98
D	1.75	44.95
E	1.75	45.23
F	1.75	52.64
G	1.75	57.08
H	1.75	57.88
I	1.75	61.31
J	1.25	53.37
K	1.31	50.06
L	2.09	28.59
M	2.56	23.02
N	2.02	42.20
P	2.2020	41.412
Q	2.4048	40.374
R	3.0	44.3
S	1.3333	54.293
T	1.397	52.118

By (6.4.8), the Mach number M_2 in the central region is given by

$$M_2^2 = \frac{1 + (\gamma - 1)M_1^2 \sin^2 \phi_1 + M_1^4 \left(\frac{1}{4}(\gamma + 1)^2 - \gamma \sin^2 \phi_1\right) \sin^2 \phi_1}{\left(1 + \frac{1}{2}(\gamma - 1)M_1^2 \sin^2 \phi_1\right)\left(\gamma M_1^2 \sin^2 \phi_1 - \frac{1}{2}(\gamma - 1)\right)}.$$

(11.3.2)

The polar for the reflected shock is given parametrically in terms of ϕ_2 by

$$\tan \theta_2 = \frac{\left(M_2^2 \sin^2 \phi_2 - 1\right) \cot \phi_2}{1 + M_2^2\left(\frac{1}{2}(\gamma + 1) - \sin^2 \phi_2\right)},$$

(11.3.3a)

$$\frac{p_3}{p_2} = 1 + \frac{2\gamma}{\gamma + 1}\left(M_2^2 \sin^2 \phi_2 - 1\right).$$

(11.3.3b)

Here M_2 is fixed at the value given by (11.3.2). The downstream Mach number

M_3 is given by

$$M_3^2 = \frac{1 + (\gamma - 1)M_2^2 \sin^2 \phi_2 + M_2^4\left(\frac{1}{4}(\gamma + 1)^2 - \sin^2 \phi_2\right)\sin^2 \phi_2}{\left(1 + \frac{1}{2}(\gamma - 1)M_2^2 \sin^2 \phi_2\right)\left(\gamma M_2^2 \sin^2 \phi_2 - \frac{1}{2}(\gamma - 1)\right)}. \quad (11.3.4)$$

Since the downstream flow must be parallel to the wall, Figure 11.2.1a shows that the deflection angles at the incident and reflected shocks are equal, that is,

$$\theta_1 = \theta_2. \quad (11.3.5)$$

Therefore the right-hand sides of (11.3.1a) and (11.3.3a) are equal. The resulting equation gives ϕ_2 in terms of M_1, ϕ_1, and M_2, and hence by (11.3.2) in terms of M_1 and ϕ_1 only. Thus all quantities are determined by M_1 and ϕ_1.

To investigate when the above procedure gives real solutions, it is convenient to superpose on one diagram the incident and reflected polars, (11.3.1) and (11.3.3), but with the vertex of the reflected polar placed on the incident polar, and with θ_2, measured from this vertex, increasing to the left. The resulting superposition is shown in Figure 11.2.2 for $M_1 = 2, 3, 4$, and 5 and $\phi_1 = 20°, 30°, 40°, 50°$, and $60°$. The dashed curves, representing incident polars, are the same as the curves shown in Figure 6.4.8b, the properties of which were noted towards the end of Section 6.4. Thus a path anticlockwise round an incident polar, starting and finishing at the vertex, corresponds to increasing ϕ_1 from the Mach angle $\sin^{-1}(1/M_1)$ up to its supplement $\pi - \sin^{-1}(1/M_1)$. The highest point on the polar corresponds to $\phi_1 = \frac{1}{2}\pi$. Regular reflection depends only on the right half of an incident polar, with $\phi_1 \leq \frac{1}{2}\pi$; but Mach reflection depends on the whole incident polar, because with the substitutions $(\phi_1, \theta_1, p_2) \to (\phi_3, \theta_3, p_4)$ the polar represents conditions at the top of the Mach stem, as shown in Figure 11.2.1b, and the range $\phi_3 \geq \frac{1}{2}\pi$ is required for the inverse Mach reflection illustrated in Figure 11.1.3b. In Figure 11.2.2, the symbols θ and p on the axes refer to a general angle and pressure. Thus each solid curve, representing a reflected polar, has its vertex at the point $(\theta, p/p_1) = (\theta_1, p_2/p_1)$ determined by (11.3.1). By our convention that θ_2 is measured to the left from this vertex, it follows from the identity $p_3/p_1 = (p_2/p_1)(p_3/p_2)$ that in the figure a reflected polar is given parametrically in terms of ϕ_2 by

$$\theta = \theta_1 - \theta_2, \quad (11.3.6a)$$

$$\frac{p}{p_1} = \left\{1 + \frac{2\gamma}{\gamma + 1}\left(M_1^2 \sin^2 \phi_1 - 1\right)\right\}\left\{1 + \frac{2\gamma}{\gamma + 1}\left(M_2^2 \sin^2 \phi_2 - 1\right)\right\}.$$

$$(11.3.6b)$$

Here θ_2 is given in terms of ϕ_2 by (11.3.3a), and M_1 and ϕ_1 are fixed. Thus M_2 is also fixed, at its value determined by (11.3.2). A path round a reflected polar, this time clockwise, starting and finishing at the vertex, corresponds to increasing ϕ_2 from the central-region Mach angle $\sin^{-1}(1/M_2)$ up to $\pi - \sin^{-1}(1/M_2)$.

11.4 Sonic Flow behind the Incident Shock

The Mach number M_2 behind the incident shock is important, because in order for there to be a reflected shock it is necessary that $M_2 \geq 1$. The contours $M_2 = 1, 2, 3,$ and 4 are shown in Figure 6.4.4b, and the contour $M_2 = 1$, representing sonic flow behind the incident shock, is redrawn as line 2 in the (M_1, ϕ_1) plane in Figure 11.2.3a. Thus shock reflection is possible only below line 2, which by (6.4.11) has equation

$$\sin^2 \phi_1 = \frac{\frac{1}{2}(\gamma - 3) + \frac{1}{2}(\gamma + 1)M_1^2 + (\gamma + 1)^{\frac{1}{2}}\left\{\frac{1}{4}(\gamma + 9) + \frac{1}{2}(\gamma - 3)M_1^2 + \frac{1}{4}(\gamma + 1)M_1^4\right\}^{\frac{1}{2}}}{2\gamma M_1^2}.$$

(11.4.1)

Correspondingly, when ϕ_1 is increased at fixed M_1 in the polar diagrams in Figure 11.2.2, the reflected polars shrink to a point as their vertices ascend to the right sonic point on the incident polar, marked as point B in Figure 6.4.7. No reflected polar can have its vertex higher than this point.

11.5 The Detachment Line for Regular Reflection

When represented in the polar diagram, the regular-reflection condition $\theta_1 = \theta_2$ becomes $\theta = 0$, so that a regular reflection corresponds to an intersection of a reflected polar with the pressure axis. The intersection point is at $p/p_1 = p_3/p_1$, and so it determines the downstream pressure p_3. Figure 11.2.2 shows that for given (M_1, ϕ_1) there are in general either two intersections or none. The lower intersection, at the smaller value of p_3, gives the "weak" or "lower-branch" regular reflection, which is the one usually observed; the upper intersection, at the larger p_3, gives the "strong" or "upper-branch" regular reflection, which might occur if there is a downstream obstruction. The values of (M_1, ϕ_1) for which the reflected polar is tangential to the pressure axis form a boundary line of the region in the (M_1, ϕ_1) plane for which regular reflection is possible. Since an increase in θ_2 corresponds to a decrease in θ, this tangency condition is satisfied when θ_2 is a maximum, say when $\theta_2 = \theta_{2,\text{max}}$, which by (6.4.19) occurs when

$$\sin^2 \phi_2 = \frac{-1 + \frac{1}{4}(\gamma + 1)M_2^2 + (\gamma + 1)^{\frac{1}{2}}\left\{1 + \frac{1}{2}(\gamma - 1)M_2^2 + \frac{1}{16}(\gamma + 1)M_2^4\right\}^{\frac{1}{2}}}{\gamma M_2^2}. \quad (11.5.1)$$

The right-hand side of (11.3.3a), with ϕ_2 given by (11.5.1) and M_2 given by (11.3.2), then becomes a function of M_1 and ϕ_1, which when equated to the right-hand side of (11.3.1a) determines line 5 in Figure 11.2.3a. Below the line, two regular reflections are possible, the weak and the strong; above, none. On the line (i.e., when $\theta_2 = \theta_{2,\max}$), the shock configuration is said to satisfy the detachment condition. The polar diagram for point E on this line, which may be called the detachment line for regular reflection, or simply the detachment line, is shown in Figure 11.2.3e. The coordinates of point E, and of other points in the figure, are given in Table 11.2.1. If a typical quantity, for example the downstream pressure p_3, is plotted vertically along an axis perpendicular to a horizontal (M_1, ϕ_1) plane, then in the resulting three-dimensional space the possible regular reflections lie on a single smoothly folded surface, on which the tangent planes are vertical along a line above the detachment line. This line separates the upper branch of the surface from the lower branch. If we imagine the surface projected vertically into the (M_1, ϕ_1) plane, we may regard the detachment line either as a sharp fold in a single-sheeted surface or as a line joining the two sheets of a double-sheeted surface.

11.6 Mach Waves

Calculations such as those in Sections 11.3–11.5 require supplementary conditions to be satisfied, to rule out complex roots of the equations. For example, the limiting case of an incident shock of zero strength (i.e., an incident Mach wave) corresponds to $\phi_1 = \sin^{-1}(1/M_1)$, that is, line 1 in Figure 11.2.3a. Only points on or above this line can give an incident shock and hence a shock reflection. As the line is approached from above, the reflected and incident polars become coincident, and $\theta_1 \to 0$. Generally, degenerate cases in which any of the shocks becomes a Mach wave are of importance in delimiting regions of the (M_1, ϕ_1) plane in which various types of configuration can occur. Note also that equations for shock angles, such as (11.4.1) for ϕ_1, or (11.5.1) for ϕ_2, have two possible solutions, with sum π, corresponding to the two halves of a polar. It is necessary to determine, graphically or otherwise, which of these angles are candidates for shock configurations. For a regular reflection we have immediately that $\phi_1 \leq \frac{1}{2}\pi$ and $\phi_2 \leq \frac{1}{2}\pi$, but a Mach reflection requires that other possibilities be considered.

11.7 The Sonic Line for Regular Reflection

When the intersection of the reflected polar with the pressure axis occurs at the sonic point of the reflected polar, the downstream flow is exactly sonic (i.e., $M_3 = 1$). By analogy with (11.4.1), this occurs when

$$
\sin^2 \phi_2 = \frac{\begin{aligned}&\tfrac{1}{2}(\gamma - 3) + \tfrac{1}{2}(\gamma + 1)M_2^2 \\ &+ (\gamma + 1)^{\frac{1}{2}}\left\{\tfrac{1}{4}(\gamma + 9) + \tfrac{1}{2}(\gamma - 3)M_2^2 + \tfrac{1}{4}(\gamma + 1)M_2^4\right\}^{\frac{1}{2}}\end{aligned}}{2\gamma M_2^2}.
$$

$$(11.7.1)$$

The corresponding values of (M_1, ϕ_1), which are found in the same way as for the detachment criterion, but with ϕ_2 given by (11.7.1) instead of (11.5.1), form line 4 in Figure 11.2.3a. On this line, the shock configuration is said to satisfy the sonic criterion for regular reflection. The polar diagram for point D on the line, which may be called the sonic line for regular reflection, or simply the sonic line, is shown in Figure 11.2.3d. Because the sonic point on a polar is below but close to the point of maximum deflection, the sonic line lies on the lower sheet of possible regular reflections, close to the detachment line at which the two sheets join up. The two lines, 4 and 5 in Figure 11.2.3a, are so close that for many purposes the distinction between them may be ignored.

11.8 Mach Reflection

We now turn to the Mach reflection, as shown in Figure 11.2.1b. Equations (11.3.1)–(11.3.4) still apply, but they need to be supplemented by the shock-polar equations applied to the Mach stem, which in terms of the Mach-stem angle ϕ_3 are

$$
\tan \theta_3 = \frac{\left(M_1^2 \sin^2 \phi_3 - 1\right) \cot \phi_3}{1 + M_1^2\left(\tfrac{1}{2}(\gamma + 1) - \sin^2 \phi_3\right)}, \tag{11.8.1a}
$$

$$
\frac{p_4}{p_1} = 1 + \frac{2\gamma}{\gamma + 1}\left(M_1^2 \sin^2 \phi_3 - 1\right). \tag{11.8.1b}
$$

Equation (11.3.5) no longer holds. Instead, continuity of flow direction and pressure across the slip line gives

$$
\theta_3 = \theta_1 - \theta_2, \qquad p_4 = p_3. \tag{11.8.2a,b}
$$

Thus $p_4/p_1 = (p_2/p_1)(p_3/p_2)$, which from (11.3.1b), (11.3.3b), and (11.8.1b) gives

$$\sin^2 \phi_3 = \frac{\frac{1}{4}(\gamma^2 - 1) + \left(\gamma M_1^2 \sin^2 \phi_1 - \frac{1}{2}(\gamma - 1)\right)\left(\gamma M_2^2 \sin^2 \phi_2 - \frac{1}{2}(\gamma - 1)\right)}{\frac{1}{2}\gamma(\gamma + 1)M_1^2}.$$

$$(11.8.3)$$

The equation $\theta_3 = \theta_1 - \theta_2$ now reduces to a single equation for ϕ_2 in terms of M_1 and ϕ_1, since in the expressions for θ_1, θ_2, and θ_3 given by (11.3.1a), (11.3.3a), and (11.8.1a) we may obtain M_2 from (11.3.2) and ϕ_3 from (11.8.3). Hence all quantities can be obtained in terms of M_1 and ϕ_1.

11.9 The Number of Mach-Reflection Configurations

The polar diagram determines when the equations in Section 11.8 have solutions. Recall that the reflected polar is given by (11.3.6) and the Mach-stem polar (11.8.1) is identical with the incident shock-polar (11.3.1), since the parameter ϕ_1 has simply been replaced by ϕ_3. Hence (11.8.2) shows that Mach reflections correspond in Figure 11.2.2 to intersections of an incident and a reflected polar (i.e., of a dashed line with a solid line). The dual aspect of a dashed curve, as representing both an incident shock and a Mach stem, implies that on it the vertex of a reflected polar is at the point with parameter ϕ_1 and the other intersections are at ϕ_3; the parameter on the reflected polar is always ϕ_2. In counting the number of intersections, it is convenient to exclude the intersection at the vertex of the reflected polar, which corresponds to a degenerate case in which the incident shock and Mach stem form a single shock on one straight line, and the reflected shock is a Mach wave. With this convention, the number of intersections, that is, the number of possible Mach-reflection configurations, can be 0, 1, 2, or 3. As (M_1, ϕ_1) is varied, the number of intersections changes by 2 when two neighbouring intersections coalesce and disappear, or conversely appear and separate. The incident and reflected polars then touch each other (i.e., have the same slope where they meet). Since the slope $(1/p_1)dp/d\theta$ at an arbitrary point on each polar is determined from (11.3.1) and (11.3.6) by appropriate forms of (6.4.32), we may equate expressions for the two slopes and express all variables in terms of M_1 and ϕ_1, as described above. The result is an equation that determines line 8c in Figure 11.2.3a. Similarly, the number of intersections changes by 1 when variation of (M_1, ϕ_1) causes an intersection point to move to the vertex of the reflected polar and disappear, or conversely emerge from this vertex and move away from it. Since the slopes of the polars are then equal at the vertex, this case is governed by the same equation that determined line 8c, but it corresponds to solutions for which $\phi_2 = \pi - \sin^{-1}(1/M_2)$.

These solutions give lines 8a and 8b,d in Figure 11.2.3a. Inspection of the polars shows that above line 8a there are no Mach-reflection configurations; between 8a and 8b,c there is 1; below 8b,d there are 2; and between 8c and 8d there are 3. The polar diagram for point H on line 8a is shown in Figure 11.2.3h. There are no Mach-reflection configurations to the left of point J, at the end of line 8a. The various configurations, like those for regular reflection, may be regarded as forming a surface comprising several branches in a three-dimensional space such as (M_1, ϕ_1, p_3) or as forming a multisheeted surface in the (M_1, ϕ_1) plane.

11.10 The von Neumann Point and Line

If a Mach reflection is such that the incident and reflected polars intersect on the pressure axis, so that $\theta_3 = 0$ and $\phi_3 = \frac{1}{2}\pi$, then the corresponding configuration can be either a regular or a Mach reflection, because the regular-reflection equation $\theta_1 = \theta_2$ and the Mach-reflection equation $\theta_3 = \theta_1 - \theta_2$ are then satisfied simultaneously. The condition $p_4 = p_3$, written again in the form $p_4/p_1 = (p_2/p_1)(p_3/p_2)$, becomes

$$\sin^2 \phi_2 = \frac{\left(1 + \frac{1}{2}(\gamma - 1)M_1^2 \sin^2 \phi_1\right)}{1 + (\gamma - 1)M_1^2 \sin^2 \phi_1 + M_1^4\left(\frac{1}{4}(\gamma + 1)^2 - \gamma \sin^2 \phi_1\right)\sin^2 \phi_1}.$$

$$\cdot \left\{M_1^2\left(\frac{1}{2}(\gamma + 1) + \frac{1}{2}(\gamma - 1)\sin^2 \phi_1\right) - \frac{1}{2}(\gamma - 1)\right\}$$

(11.10.1)

The equation $\theta_1 = \theta_2$, with M_2 given by (11.3.2) and ϕ_2 by (11.10.1), becomes a single equation in M_1 and ϕ_1, satisfied by points on line 3 in Figure 11.2.3a, the von Neumann line, on which the shock configuration is said to satisfy the von Neumann criterion. This line touches the detachment line for regular reflection at point P in Figures 11.2.3a,b, the von Neumann point. The polar diagrams for points on the von Neumann line take different forms to the left and right of the von Neumann point. To the left, the intersection of incident polar, reflected polar, and pressure axis is above the maximum-deflection point of the reflected polar and so corresponds to a strong regular reflection, which as we have noted is not usually observed unless special conditions are imposed downstream. For example, the polar diagram for point C in Figure 11.2.3a is as shown in Figure 11.2.3c. To the right, the intersection is below the maximum-deflection point and so corresponds to a weak regular reflection. For example, the polar diagram for $(M_1, \phi_1) \simeq (5.0, 31°)$ is of this type, obtainable from Figure 11.2.2d by raising slightly the reflected polar for $\phi_1 = 30°$. Exactly at the von Neumann point, the intersection is at the maximum-deflection point of the reflected polar, where the tangent is therefore vertical, and so the polars are as in Figure 11.2.3e, but with the point of tangency moved up to the dashed line.

The defining property of the von Neumann line, that the corresponding Mach reflection satisfies the von Neumann criterion $\theta_3 = 0$ and $\phi_3 = \frac{1}{2}\pi$, is simply that the Mach stem is a normal shock perpendicular to the wall, as shown in Figure 11.1.3a. The Mach stem can be any length, and the slip line is parallel to the wall. In the limit as the length of the Mach stem tends to zero, the shock pattern becomes a regular reflection. Inspection of Figure 11.2.2 shows that, at fixed M_1, as ϕ_1 increases through the von Neumann line so θ_3 increases through zero, since the intersection point of the incident and reflected polars moves from left to right across the pressure axis. Therefore $\theta_3 < 0$ below the von Neumann line, and $\theta_3 > 0$ above it. That is, in the (M_1, ϕ_1) plane the region below the von Neumann line corresponds to inverse Mach reflection, as in Figure 11.1.3b, and the region above to direct Mach reflection, as in Figure 11.2.1b.

11.11 Henderson Points, Lines, and Regions.
The von Neumann Region

If a Mach reflection is such that the incident and reflected polars intersect at a sonic point on the reflected polar (i.e., where $M_3 = 1$), then the flow downstream of the reflected shock is exactly sonic. The corresponding values of (M_1, ϕ_1) are found from the Mach-reflection equation $\theta_3 = \theta_1 - \theta_2$, with M_2 given by (11.3.2), ϕ_2 by (11.7.1), and ϕ_3 by (11.8.3). The left sonic point, with $\phi_2 < \frac{1}{2}\pi$, and the right sonic point, with $\phi_2 > \frac{1}{2}\pi$, may be considered separately, as may the direct Mach reflection, with $\phi_3 \leq \frac{1}{2}\pi$, and the inverse Mach reflection, with $\phi_3 \geq \frac{1}{2}\pi$. Since the right half of a reflected polar does not intersect the left half of an incident polar, which we have seen is also the Mach-stem polar, it follows that the combination (right sonic point, inverse Mach reflection) is impossible. The combination (right sonic point, direct Mach reflection) gives line 7a in Figure 11.2.3a, the first Henderson line, which intersects line 4, the sonic line for regular reflection, at point T, the first Henderson point. The polar diagram for point G on line 7a is shown in Figure 11.2.3g. The combinations (left sonic point, direct Mach reflection) and (left sonic point, inverse Mach reflection) give lines 7b and 7c respectively, together forming line 7b,c, the second Henderson line, which intersects line 4 and line 3, the von Neumann line, at point Q, the second Henderson point. For points on the Henderson lines, the corresponding shock configurations are said to satisfy the sonic criterion for Mach reflection. Inspection of Figure 11.2.2 shows that, in Mach reflection, $M_3 < 1$ for points (M_1, ϕ_1) between the two Henderson lines, but $M_3 > 1$ elsewhere. We have seen that Mach reflection is possible only below line 8a in Figure 11.2.3a. Therefore in Mach reflection the flow downstream of the reflected shock is supersonic for points (M_1, ϕ_1) in two separate regions, one above the first Henderson line but below line 8a, the first Henderson region, and the other to the right of the second Henderson line, the second Henderson region. These two regions are of great interest, because a downstream supersonic flow can support further wave patterns and shocks.

In the first Henderson region and above are values of (M_1, ϕ_1) for which there occurs in experiments a regular or Mach reflection that on our algebraic theory, based on shock polars, is impossible (i.e., for which the "von Neumann paradox" occurs). Transition is then often to the von Neumann reflection shown in Figure 11.1.3h, although this configuration can be difficult to distinguish from a Mach reflection when the fan-shaped compression region is thin. Thus an apparently "impossible" Mach reflection may in fact be a von Neumann reflection.

The region of the (M_1, ϕ_1) plane above line 8a but below line 2 and to the right of line 1 may be called the von Neumann region, even though it does not correspond closely to the region of von Neumann reflection. The von Neumann paradox occurs in some other regions and strips of the (M_1, ϕ_1) plane besides the first Henderson region and the von Neumann region. A well-known example of the paradox is the persistence of regular reflection up to "impossible" values of ϕ_1 when a shock is diffracted by a wedge or by a concave corner in a wall. This has been attributed to viscous effects and thus lies outside the scope of our inviscid analysis. In the second Henderson region, the reflected shock can develop reversed curvature next to a fan-shaped compression wave, as in the transitional Mach reflection shown in Figure 11.1.3f, or can develop a kink forming the triple point of a second Mach reflection, as in the double Mach reflection shown in Figure 11.1.3g; and there are many other possibilities.

Incident shocks with M_1 less than its value at the second Henderson point have been called weak, and those with greater M_1 have been called strong. But terminology is not uniform, and the dividing point between weak and strong incident shocks has sometimes been taken to be the value of M_1 at the von Neumann point. Incident shocks with M_1 less than its value at the first Henderson point have been called very weak, and those with M_1 less than its value at point S in Figure 11.2.3a have been called extremely weak. These terms are now used only infrequently, but they appear in earlier writing.

11.12 The Normal Reflected-Shock Line for Mach Reflection

If a Mach reflection is such that the incident and reflected polars intersect at the top of the reflected polar, so that $\theta_2 = 0$ and $\phi_2 = \frac{1}{2}\pi$, then the reflected shock is a normal shock. The corresponding values of (M_1, ϕ_1) are found from the equation $\theta_3 = \theta_1$, in which ϕ_3 is given by (11.8.3) with $\sin \phi_2 = 1$ and with

M_2 obtained from (11.3.2). The result is line 6 in Figure 11.2.3a, the normal reflected-shock line for Mach reflection. Since the vertex of the reflected polar is on the right half of the incident polar, so also is the intersection point at the top. Therefore $\theta_3 \geq 0$ and $\phi_3 \leq \frac{1}{2}\pi$, so that the configuration is a direct Mach reflection, consistent with the fact that the normal reflected-shock line is above the von Neumann line. The two lines intersect on the line for a Mach-wave incident shock, that is, at point A in Figure 11.2.3a, where $\theta_1 = \theta_2 = \theta_3 = 0$ and $\phi_2 = \phi_3 = \frac{1}{2}\pi$. Hence point A corresponds to a degenerate configuration in which the Mach stem and reflected shock form a single normal shock, perpendicular to the wall, and the incident shock is a Mach wave. The polar diagram for point F on the normal reflected-shock line is shown in Figure 11.2.3f. Inspection of Figures 11.2.2 and 11.2.3e–g shows that, at fixed M_1, as ϕ_1 increases through the normal reflected-shock line, so the intersection point of the incident and reflected polars moves to the right on the reflected polar, which with our sign convention for θ_2 implies that θ_2 decreases through zero. Therefore $\theta_2 < 0$ above the normal reflected-shock line, and $\theta_2 > 0$ below it. That is, in the (M_1, ϕ_1) plane the region above the normal reflected-shock line corresponds to the flow in the central region of a Mach reflection being deflected further towards the wall at the reflected shock, and the region below the flow being deflected so that it points less towards the wall.

11.13 Angles of Incidence and Reflection

The results above, and their consequences, are often expressed in other variables, for instance incident-shock Mach number $M_i = M_1 \sin \phi_1$ or incident-shock inverse pressure ratio ξ defined by

$$\xi = \frac{p_1}{p_2} = \left\{ 1 + \frac{2\gamma}{\gamma + 1} \left(M_1^2 \sin^2 \phi_1 - 1 \right) \right\}^{-1}. \qquad (11.13.1)$$

For example, the dependence of angle of reflection $\phi_2 - \theta_1$ on angle of incidence ϕ_1 in a lower-branch regular reflection is shown in Figure 11.13.1 as a series of curves at fixed p_2/p_1, equivalent to fixed ξ. The maximum angle of incidence for regular reflection at given p_2/p_1 is attained at the dashed line in the figure. The curves with $p_2/p_1 \leq 7.02$ pass through the point for which the angles of incidence and reflection are both equal to $39.23°$. The figure shows that the angle of reflection is less than the angle of incidence when the angle of reflection is less than $39.23°$, and conversely when it is greater. For the larger values of p_2/p_1 (i.e., for stronger shocks), the angle of reflection can be much less than the angle of incidence. In the limit $p_2/p_1 \to 1$, that is, as the incident

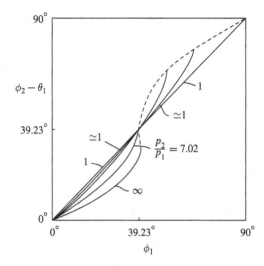

Fig. 11.13.1. Dependence of angle of reflection $\phi_2 - \theta_1$ on angle of incidence ϕ_1 for lower-branch regular reflection in a polytropic gas with $\gamma = 1.4$, at fixed values of incident pressure ratio $p_2/p_1 = 1 + (2\gamma/(\gamma + 1))(M_1^2 \sin^2 \phi_1 - 1)$. The maximum value of ϕ_1 for regular reflection is at the dashed line, on our simplified theory. The curves for $p_2/p_1 \leq 7.02$ pass through the point $(\phi_1, \phi_2 - \theta_1) = (39.23°, 39.23°)$.

shock becomes a Mach wave, the curves become straight lines and so give the standard result that in acoustics the angle of incidence equals the angle of reflection. The dependence of angle of reflection on angle of incidence may likewise be calculated for Mach reflection.

11.14 Curved Shocks, Effective Nozzle Flow, and Hysteresis

The theory we have presented has been extended in many ways. For example, we have considered only a small enough neighbourhood of the Mach triple point that the Mach stem may be assumed to be a straight line, so that the Mach-stem angle ϕ_3 and the corresponding deflection angle θ_3 are constant. In fact, the whole Mach stem, curved to meet the wall at right angles, is represented in a polar diagram by the high arc on the Mach-stem polar between the points of intersection of the polar with the reflected polar and with the pressure axis. Recall that the Mach-stem polar is the same as the incident polar, but with the parameter ϕ_1 renamed ϕ_3, and the top of the polar, where it intersects the pressure axis, corresponds to a normal shock and no deflection. In the same way, a curved reflected shock is represented by an arc on the reflected polar. Multiple-shock patterns, and other flow patterns as shown in Figure 11.1.3, admit polar representations of some sophistication.

Other extensions of the theory depend on analysis of conditions downstream. In Figure 11.2.1b, the flow in region 4, that is, downstream of the Mach stem, between the wall and the slip line, is effectively a nozzle flow and may be analysed by the methods of Sections 7.2 and 7.3. The shape of the slip line is strongly affected by whether it intercepts any wave patterns, for example a Prandtl–Meyer expansion fan, propagating from the rear of the device or structure that produced the incident shock. Thus the region downstream of the Mach stem may at first decrease in width and then expand, so that the flow resembles that in a Laval nozzle and can become choked. Hence a global analysis is possible of some flow fields containing a Mach reflection, and the analysis can lead to a prediction of the height of the Mach stem. One aspect of such an analysis is the direction of "information flow" on a shock, since this determines appropriate boundary conditions. For example, at a direct Mach configuration, as shown in Figure 11.2.1b, an incoming incident shock splits into three outgoing parts: the reflected shock, the Mach stem, and the slip line; but at an inverse Mach reflection, as shown in Figure 11.1.3b, the incoming parts are the incident shock and the Mach stem, and only the reflected shock and the slip line are outgoing.

At the end of Section 11.1 we said a little about experimental results on transitions between different shock configurations. Expressed in terms of Figure 11.2.3a, our remarks related to the triangular region bounded by lines 3 and 4 to the right of point Q: In steady flow, a transition from regular to Mach reflection often occurs on passing upwards through line 3, the von Neumann line, and in pseudo-steady flow the transition often occurs on line 4, the sonic line for regular reflection. Detailed experimental results, including their dependence on the whole range of (M_1, ϕ_1) and on many other parameters besides, are the subject of numerous research papers and at least two books, but they lie beyond the scope of this book. Hysteresis is discussed in much of this work. The reader will find that the various points, lines, and regions we have defined in terms of M_1 and ϕ_1 provide an invaluable aid to understanding the varied phenomena that occur. Thus the (M_1, ϕ_1) plane, in Figure 11.2.3a, provides a fine example of the way in which a suitable parameter space can render a complex subject matter intelligible.

11.15 Possible and Impossible Configurations

The von Neumann–Henderson theory presented in Sections 11.2–11.14 is an example of a more general theory of possible and impossible configurations. Let us take a point in a plane and from it draw a number of straight lines, dividing the region around the point into sectors. Assume that each line is a shock, a Mach line, a slip line, or a streamline, and assume that each sector is a region of uniform rectilinear flow, a fan region of a Prandtl–Meyer expansion, a "dead" region of no flow, or a solid body. Then any combination of the different types of lines and sectors can be analysed and classified as always possible, always impossible, or possible only for some angles, shock strengths, and flow speeds. A generic configuration is shown in Figure 11.15.1a, and a selection of possible and impossible configurations is shown in Figure 11.15.1b–e. In many problems, the appropriate configuration is not obvious in advance. For example, in the refraction of a Mach line by a shock, the configuration contains, in addition to the refracted Mach line, a streamline across which the vorticity and entropy are not smooth, as shown in Figure 11.15.1b.

One of the simplest impossible configurations is a triple shock intersection with no other structure. If such a configuration is hypothesized, as in Figure 11.15.1c, the jump conditions imply that the shocks all have zero strength or that one of them has zero strength and the other two have the same strength and lie on one straight line (i.e., that in effect there is only one shock). Therefore if three shocks meet at a point, then from the point there emerges another line or sector, usually a slip line. This result follows from a general theorem, called the triple-shock entropy theorem, which applies to any fluid with a convex equation of state.

We have already met several of the more complicated types of configuration. For example, the reflection of a shock off the edge of a jet, as shown in Figure 11.15.1d, can produce a configuration with five lines – two slip lines, two Mach lines, and a shock – and five sectors, of which three are regions of uniform rectilinear flow, one is a dead region of no flow, and one is a Prandtl–Meyer expansion fan. We saw the configuration in an overexpanded jet emerging from a Laval nozzle in Figure 7.3.2a,b.

Another configuration with five lines and five sectors occurs in the coalescence of two shocks from the same family, for example in the supersonic flow next to a concave polygonal wall with two corners. Two attached oblique shocks, one from each corner, can coalesce to form a single shock, and from the point of coalescence there emerge a slip line and a Mach line, as shown in Figure 11.15.1e. The flow at each corner, with its attached oblique shock, also forms a configuration within our scheme, containing three lines – two streamlines and a shock – and three sectors, two containing uniform rectilinear flow and one consisting of a solid body (i.e., the wall). Another configuration is produced where the Mach line from the point of coalescence of the shocks meets the wall and is reflected; and then another where the reflected Mach line meets the slip line from the coalescence; and so on. Many other high speed flows contain several configurations of the generic type shown in Figure 11.15.1a, and their properties have been well studied.

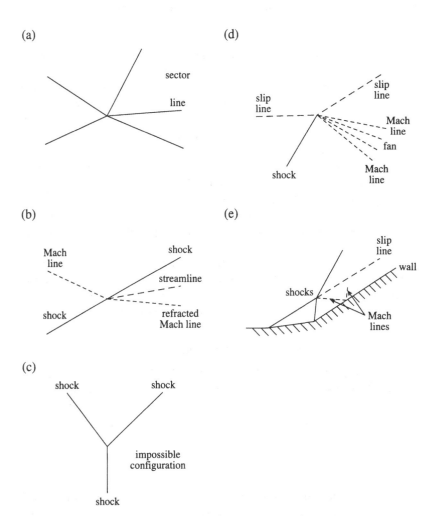

Fig. 11.15.1. Possible and impossible configurations. (a) Generic configuration of lines and sectors. (b) Refraction of a Mach line by a shock. Vorticity and entropy are not smooth across the streamline shown. (c) An impossible configuration. If three shocks meet at a point, there must be a slip line or another sector. (d) Reflection of a shock, as a Prandtl–Meyer expansion, off a constant pressure boundary, for example off the edge of a jet. (e) Coalescence of shocks of the same family, originating as oblique attached shocks at the corners of a concave polygonal wall. The figure contains four configurations of generic type (a).

11.16 Bibliographic Notes

The earliest observations of regular and Mach reflection are described in Mach (1878). The theory begins with von Neumann (1943) and continues with, for example, Henderson (1964, 1965, 1987), Ben-Dor & Glass (1979, 1980), Ben-Dor & Takayama (1985), and Ben-Dor (1987). Two influential papers containing experimental results were Henderson & Lozzi (1975, 1979). An early reference article on shock reflections and intersections is Polachek & Seeger (1958), and the main texts and monographs, covering both theory and experiment, are Courant & Friedrichs (1948), Ben-Dor (1992), and Glass & Sislian (1994); three surveys and reviews are Lesser & Field (1983), Hornung (1986), and Ben-Dor (1988). Research papers in Table 13.6.1a on shock reflections and intersections are Colella & Henderson (1990), Grove & Menikoff (1990), Barkhudarov et al. (1991), Henderson, Colella & Puckett (1991), Sasoh & Takayama (1994), Chpoun, Passerel, Li & Ben-Dor (1995), Vuillon, Zeitoun & Ben-Dor (1995), Buttsworth (1996), Liu (1996), Henderson, Crutchfield & Virgona (1997), Li & Ben-Dor (1997a, b), Olejniczak, Wright & Candler (1997), and Henderson & Menikoff (1998). See also Skews (1997) and Hornung (1998). The subject of shock intersections overlaps with that of shock focusing; see, for example, the photographs in Sturtevant & Kulkarny (1976). The curvature of reflected shocks and Mach stems is closely related to their generation of vorticity, as discussed in Kevlahan (1997). Photographs of shock reflections and intersections appear in Van Dyke (1982). The interaction of a shock with a boundary layer forms a large subject in its own right, to which an important early contribution was Lighthill (1953b).

11.17 Further Results and Exercises

1. At an oblique shock in a polytropic gas with ratio of specific heats γ, the normal components of the velocity of the gas on the two sides of the shock are u_{1n} and u_{2n}, the tangential component of velocity is u_t, and the critical speed is $\hat{c}$. The quantity λ is defined by $\lambda^2 = (\gamma - 1)/(\gamma + 1)$. Obtain Prandtl's relation in the form

$$u_{1n}u_{2n} + \lambda^2 u_t^2 = \hat{c}^2.$$

A stationary oblique shock in a uniform supersonic stream with critical speed $\hat{c}$ is attached at an angle ϕ to a plane wall. Obtain equations sufficient in general to determine the angle between the wall and the reflected shock.

2. Obtain the normal shock relations in the form

$$\frac{p_2}{p_1} = 1 + \frac{2\gamma}{\gamma + 1}(M_1^2 - 1),$$

$$\frac{\rho_2}{\rho_1} = \frac{\frac{1}{2}(\gamma + 1)M_1^2}{1 + \frac{1}{2}(\gamma - 1)M_1^2},$$

$$M_2^2 = \frac{1 + \frac{1}{2}(\gamma - 1)M_1^2}{\gamma M_1^2 - \frac{1}{2}(\gamma - 1)}.$$

Show how these may be transformed to give the corresponding oblique shock relations. Assuming that the flow deflection angle θ is small, and that all calculations may be performed only to the first order in θ, obtain an expression for the pressure ratio across an oblique shock that has been reflected regularly at a plane wall. Express your answer in terms of the upstream pressure p_1 and upstream Mach number M_1, and show that the downstream Mach number is

$$M_1 - \left(1 + \frac{1}{2}(\gamma - 1)M_1^2\right)\frac{2M_1\theta}{(M_1^2 - 1)^{\frac{1}{2}}}.$$

3. Obtain numerically, for all branches of regular and Mach reflection in a polytropic gas with ratio of specific heats $\gamma = 1.4$, the downstream pressure p_3 and downstream Mach number M_3 behind the reflected shock as functions of incident Mach number M_1 and incident shock angle ϕ_1. Present your results as two-dimensional sections, at fixed M_1, of surfaces in the three-dimensional spaces (M_1, ϕ_1, p_3) or (M_1, ϕ_1, M_3), and include the sections $M_1 = 1.36, 1.44, 1.75, 2.3$, and 3.5. In each section, the branches are represented by superposed graphs of p_3 or M_3 as functions of ϕ_1. Use these superposed graphs to discuss possible hysteresis effects that might occur as ϕ_1 is varied at fixed M_1. Relate your discussion to the lines shown in Figure 11.2.3a.

12

The Hodograph Method

12.1 Arbitrary Fluid

A technique for analysing steady two-dimensional irrotational homentropic flow of an ideal fluid is to regard position $\mathbf{x} = (x, y)$ as a function of velocity $\mathbf{u} = (u, v)$ and determine the partial differential equations satisfied by $\mathbf{x}(\mathbf{u})$. The technique is the hodograph method, and the transformation to independent variables (u, v) and dependent variables (x, y) is a hodograph transformation. The point of the method is that the equations satisfied by $\mathbf{x}(\mathbf{u})$ (i.e., the hodograph equations) are linear in $\mathbf{x}$ and have solutions readily available by standard techniques. Inversion of $\mathbf{x}(\mathbf{u})$, where this is possible, gives a solution $\mathbf{u}(\mathbf{x})$ of the original nonlinear equations in $\mathbf{u}$. A disadvantage of the method is that it renders the boundary conditions nonlinear, so that only rarely can it be used to solve completely a boundary-value problem. Particular regions in complicated flows can nevertheless be usefully analysed by the hodograph method, and it has been well studied.

Let position in a plane be denoted by the complex variable z. At any point, the fluid density is ρ, the flow speed is q, and the flow direction is at an angle θ to the x axis. Thus the complex variables z and $qe^{i\theta}$ are defined in terms of (x, y) and (u, v) by

$$z = x + iy, \qquad qe^{i\theta} = u + iv. \tag{12.1.1}$$

Partial differentiation will be denoted by a subscript. Then the equations expressing irrotationality and conservation of mass are

$$v_x - u_y = 0, \qquad (\rho u)_x + (\rho v)_y = 0. \tag{12.1.2}$$

In homentropic flow, the thermodynamic variables apart from entropy are all functions of any one of them. Since Bernoulli's equation, in terms of enthalpy $h = \int \rho^{-1} \, dp$, is $h + \frac{1}{2}q^2 = $ constant, it follows that h, p, and ρ may be

206

regarded as functions of q. Let us select a reference density ρ_0, and define the velocity potential ϕ and stream function ψ to satisfy the equations

$$(u, v) = (\phi_x, \phi_y), \qquad (\rho u, \rho v) = (\rho_0 \psi_y, -\rho_0 \psi_x). \qquad (12.1.3)$$

Then

$$d\phi = \phi_x dx + \phi_y dy = u\, dx + v\, dy,$$
$$\frac{\rho_0}{\rho} d\psi = \frac{\rho_0}{\rho} \psi_x dx + \frac{\rho_0}{\rho} \psi_y dy = -v\, dx + u\, dy. \qquad (12.1.4)$$

Therefore

$$d\phi + \frac{i\rho_0}{\rho} d\psi = (u - iv)(dx + i\, dy) = q e^{-i\theta}\, dz. \qquad (12.1.5)$$

Hence

$$dz = \frac{e^{i\theta}}{q}\left(d\phi + \frac{i\rho_0}{\rho} d\psi\right). \qquad (12.1.6)$$

To obtain the hodograph equations, we regard z, ϕ, and ψ in (12.1.6) as functions of q and θ. Thus we write dz, $d\phi$, and $d\psi$ in terms of dq and $d\theta$ as

$$dz = z_q dq + z_\theta d\theta,$$
$$d\phi = \phi_q dq + \phi_\theta d\theta, \qquad (12.1.7)$$
$$d\psi = \psi_q dq + \psi_\theta d\theta.$$

Since q and θ are independent, the resulting coefficients of dq and $d\theta$ in (12.1.6) give separate equations. Thus

$$(z_q, z_\theta) = \frac{e^{i\theta}}{q}\left(\phi_q + \frac{i\rho_0}{\rho} \psi_q, \phi_\theta + \frac{i\rho_0}{\rho} \psi_\theta\right). \qquad (12.1.8)$$

Equality of the mixed derivatives $z_{q\theta}$ and $z_{\theta q}$ gives

$$\left(\frac{e^{i\theta}}{q}\left(\phi_q + \frac{i\rho_0}{\rho} \psi_q\right)\right)_\theta = \left(\frac{e^{i\theta}}{q}\left(\phi_\theta + \frac{i\rho_0}{\rho} \psi_\theta\right)\right)_q. \qquad (12.1.9)$$

Recall that ρ depends only on q. Hence evaluating the partial derivatives, cancelling terms, and dividing by $e^{i\theta}$, we obtain

$$\frac{i}{q}\left(\phi_q + \frac{i\rho_0}{\rho} \psi_q\right) = -\frac{\phi_\theta}{q^2} + i\rho_0\left(\frac{1}{\rho q}\right)_q \psi_\theta. \qquad (12.1.10)$$

The imaginary and real parts of this equation give

$$\phi_q = \left(\frac{\rho_0}{\rho q}\right)_q q\psi_\theta, \qquad \phi_\theta = \frac{\rho_0}{\rho}q\psi_q. \qquad (12.1.11)$$

Since ρ is a known function of q, Equations (12.1.11) form a pair of simultaneous equations for the potential ϕ and stream function ψ. These are the hodograph equations. They are linear and may be written in several equivalent ways, for example as a single second-order equation in ϕ or ψ. To obtain the equation in ψ, we equate the mixed derivatives $\phi_{q\theta}$ and $\phi_{\theta q}$. Thus

$$\left(\left(\frac{\rho_0}{\rho q}\right)_q q\psi_\theta\right)_\theta = \left(\frac{\rho_0}{\rho}q\psi_q\right)_q. \qquad (12.1.12)$$

Therefore

$$\left(\frac{q}{\rho}\right)\psi_{qq} + \left(\frac{q}{\rho}\right)_q \psi_q - \left(\frac{1}{\rho q}\right)_q q\psi_{\theta\theta} = 0. \qquad (12.1.13)$$

Let the speed of sound in the fluid be c, so that the Mach number is $M = q/c$. Bernoulli's equation $\int \rho^{-1}\,dp + \frac{1}{2}q^2 = $ constant is equivalent to $\rho^{-1}\,dp + qdq = 0$, that is, $dp/dq = -\rho q$. Division by $dp/d\rho = c^2$ gives

$$\frac{d\rho}{dq} = -\frac{\rho q}{c^2} = -M^2\frac{\rho}{q}. \qquad (12.1.14)$$

Therefore evaluation of the coefficients in (12.1.13) gives

$$q^2\psi_{qq} + (1 + M^2)q\psi_q + (1 - M^2)\psi_{\theta\theta} = 0. \qquad (12.1.15)$$

In terms of the operator qd/dq, this is

$$\left(q\frac{d}{dq}\right)^2\psi + M^2\left(q\frac{d}{dq}\right)\psi + (1 - M^2)\psi_{\theta\theta} = 0. \qquad (12.1.16)$$

Equation (12.1.15) or (12.1.16) for the stream function ψ is Chaplygin's equation. We shall later obtain several families of solutions, starting from (12.1.11) written in terms of M and qd/dq, that is,

$$q\phi_q = -\frac{\rho_0}{\rho}(1 - M^2)\psi_\theta, \qquad \phi_\theta = \frac{\rho_0}{\rho}q\psi_q. \qquad (12.1.17)$$

In the above equations, M depends on q.

Chaplygin's equation is elliptic when $M < 1$ and hyperbolic when $M > 1$ (i.e., is of mixed type). Near a curve on which $M = 1$ (i.e., near a sonic line), we may use instead of q a variable η proportional to $M - 1$, and instead of ψ a variable Ψ that combines the terms in ψ_{qq} and ψ_q into one term, and hence transform Chaplygin's equation locally into the standard equation of mixed type (i.e., Tricomi's equation $\Psi_{\eta\eta} - \eta\Psi_{\theta\theta} = 0$). Thus in the hodograph plane, the study of flow near a sonic line reduces to analysis of solutions of Tricomi's equation, for which there is an extensive theory.

12.2 Polytropic Gas

For a polytropic gas, with ratio of specific heats γ, we can obtain expressions for $\rho, c, M, \ldots$ in terms of q and γ. The stagnation condition, at which $q = 0$, will be indicated by a subscript 0, so that the stagnation sound speed is c_0; the vacuum condition, at which q is a maximum and $c = 0$, will be indicated by a subscript m, so that the maximum flow speed is q_m; and the critical condition, at which $q = c$ (i.e., at which the flow is sonic) will be indicated by the symbol $\hat{}$, so that the critical speed and sound speed are $\hat{c}$. Then Bernoulli's equation is

$$\frac{c^2}{\gamma - 1} + \frac{q^2}{2} = \frac{c_0^2}{\gamma - 1} = \frac{q_m^2}{2} = \frac{1}{2}\left(\frac{\gamma + 1}{\gamma - 1}\right)\hat{c}^2. \tag{12.2.1}$$

Dividing by the constants on the right gives

$$\left(\frac{c}{c_0}\right)^2 + \left(\frac{q}{q_m}\right)^2 = 1. \tag{12.2.2}$$

Let the stagnation density be ρ_0, and write the homentropic relation $\rho/\rho_0 = (c/c_0)^{2/(\gamma-1)}$ in terms of the quantity $\beta = 1/(\gamma - 1)$ as

$$\left(\frac{c}{c_0}\right)^2 = \left(\frac{\rho}{\rho_0}\right)^{\frac{1}{\beta}}. \tag{12.2.3}$$

For a diatomic gas, with $\gamma = 7/5$, we have $\beta = 5/2$. Define τ by

$$\tau = \left(\frac{q}{q_m}\right)^2. \tag{12.2.4}$$

Then (12.2.2) is

$$\left(\frac{\rho}{\rho_0}\right)^{\frac{1}{\beta}} + \tau = 1, \tag{12.2.5}$$

or

$$\frac{\rho}{\rho_0} = (1 - \tau)^{\beta}. \tag{12.2.6}$$

We now express M in terms of τ. Division of Bernoulli's equation (12.2.1), with right-hand side $\frac{1}{2}q_{\mathrm{m}}^2$, by $\frac{1}{2}q^2$ gives

$$\left(\frac{2}{\gamma - 1}\right)\frac{1}{M^2} + 1 = \frac{1}{\tau}, \tag{12.2.7}$$

or

$$M^2 = \frac{2\beta\tau}{1 - \tau}. \tag{12.2.8}$$

Equations (12.2.6) and (12.2.8), though unfamiliar in appearance, are simply Bernoulli's equation in new variables and show that ρ and M take simple forms when written in terms of τ. Henceforth, τ and θ will be taken as the independent variables, so that in the hodograph equations we put

$$q\frac{\partial}{\partial q} = 2\tau\frac{\partial}{\partial \tau}. \tag{12.2.9}$$

Then with q from (12.2.4), M^2 from (12.2.8), and $q\partial/\partial q$ from (12.2.9), the hodograph equations (12.1.17) become

$$2\tau(1 - \tau)^{\beta+1}\phi_\tau = -(1 - (2\beta + 1)\tau)\psi_\theta, \qquad (1 - \tau)^\beta\phi_\theta = 2\tau\psi_\tau. \tag{12.2.10}$$

Chaplygin's equation (12.1.16) becomes

$$4\tau^2(1 - \tau)\psi_{\tau\tau} + 4(1 + (\beta - 1)\tau)\tau\psi_\tau + (1 - (2\beta + 1)\tau)\psi_{\theta\theta} = 0. \tag{12.2.11}$$

We now investigate a family of separable solutions of Chaplygin's equation, containing an integer parameter m and an arbitrary constant ϵ_m, by determining solutions with θ dependence $\sin(m\theta + \epsilon_m)$. This is a natural family to consider, because θ is an angle, and it leads to physically realistic solutions in some flow regions. It is convenient to write the τ dependence as $\tau^{\frac{1}{2}m}f(\tau)$. Thus we look for stream functions ψ of the form

$$\psi = \tau^{\frac{1}{2}m}f(\tau)\sin(m\theta + \epsilon_m). \tag{12.2.12}$$

Substitution into (12.2.11) shows that $f(\tau)$ must satisfy the equation

$$\tau(1 - \tau)f'' + ((m + 1) - (m + 1 - \beta)\tau)f' + \tfrac{1}{2}m(m + 1)\beta f = 0. \tag{12.2.13}$$

The hypergeometric equation for a function $y(x)$ depending on parameters r, s, and t is

$$x(1 - x)y'' + (t - (r + s + 1)x)y' - rsy = 0, \tag{12.2.14}$$

and one solution is the hypergeometric function $F(r, s, t; x)$, defined by its Taylor series expansion as

$$
\begin{aligned}
y(x) &= F(r, s, t; x) \\
&= 1 + \frac{r \cdot s}{t \cdot 1} x + \frac{r(r+1) \cdot s(s+1)}{t(t+1) \cdot 1 \cdot 2} x^2 + \cdots .
\end{aligned}
\tag{12.2.15}
$$

Equations (12.2.13) and (12.2.14) are equivalent if

$$
r + s = m - \beta, \qquad rs = -\tfrac{1}{2}m(m+1)\beta, \qquad t = m + 1. \tag{12.2.16}
$$

Thus we obtain a family of stream functions ψ satisfying Chaplygin's equation. The corresponding potentials ϕ are obtained by substituting the functional form (12.2.12) for ψ into the hodograph equations (12.2.10) for ϕ and ψ. The result, when $\epsilon_m = 0$, is

$$
\psi = \tau^{\frac{1}{2}m} f(\tau) \sin m\theta,
$$
$$
\phi = -\tau^{\frac{1}{2}m}(1 - \tau)^{-\beta} \left(f(\tau) + \frac{2\tau}{m} f'(\tau) \right) \cos m\theta. \tag{12.2.17}
$$

Therefore a family of solutions of the hodograph equations, and hence of the original fluid dynamical equations, can be investigated using properties of hypergeometric functions. By the linearity of the hodograph equations, we may take arbitrary linear combinations, including infinite sums and integrals, of these solutions, to obtain further solutions.

Three simple types of separable solution of the hodograph equations, obtainable directly or as limits of hypergeometric solutions, are (a) $\psi \propto \theta, \phi = f(\tau)$, in which the streamlines $\psi = $ constant are lines of constant θ, representing a source or sink; (b) $\psi = f(\tau), \phi \propto \theta$, in which the potential lines $\phi = $ constant are lines of constant θ, representing a vortex; and (c) an arbitrary linear combination of (a) and (b) (i.e., a spiral flow). Even these solutions reveal a difficulty of the hodograph method, namely lack of global invertibility in the transformation between position coordinates and velocity components. We discuss this difficulty in the next two sections.

12.3 Ringleb's Flow

To illustrate the type of flow of a polytropic gas considered in the previous section, we shall consider a flow in which the θ dependence of ψ is $\sin \theta$. Then either directly from the hodograph equations (12.2.10), or from the hypergeometric solutions (12.2.12)–(12.2.17) with $m = 1$ and $\epsilon_m = 0$, we obtain a solution

$$\psi = \tau^{-\frac{1}{2}} \sin \theta, \qquad (12.3.1a)$$

$$\phi = \tau^{-\frac{1}{2}}(1 - \tau)^{-\beta} \cos \theta. \qquad (12.3.1b)$$

This solution gives Ringleb's flow. Starting from (12.3.1) we shall deduce its properties in the physical plane, that is, the (x, y) plane.

Expressions defined in the hodograph plane may be converted to expressions in the physical plane by means of Equation (12.1.6) for dz in terms of $d\phi$ and $d\psi$. With $\rho_0/\rho = (1 - \tau)^{-\beta}$, and units such that $q_m = 1$ (i.e., $q = \tau^{\frac{1}{2}}$), we have

$$dz = \tau^{-\frac{1}{2}} e^{i\theta} (d\phi + i(1 - \tau)^{-\beta} d\psi). \qquad (12.3.2)$$

That is,

$$\begin{aligned}
z_\tau &= \tau^{-\frac{1}{2}} e^{i\theta} (\phi_\tau + i(1 - \tau)^{-\beta} \psi_\tau), \\
z_\theta &= \tau^{-\frac{1}{2}} e^{i\theta} (\phi_\theta + i(1 - \tau)^{-\beta} \psi_\theta).
\end{aligned} \qquad (12.3.3)$$

The expressions for ϕ and ψ in (12.3.1) give functions of τ and θ on the right of (12.3.3), which may thus be integrated to give z as a known function of τ and θ. It is convenient to define functions $X(\tau)$ and $R(\tau)$ by

$$X(\tau) = \frac{1}{2}\beta \int^\tau t^{-1}(1 - t)^{-\beta-1} dt, \qquad R(\tau) = \frac{1}{2}\tau^{-1}(1 - \tau)^{-\beta}. \quad (12.3.4)$$

The solution of (12.3.3) is then

$$z = X(\tau) + R(\tau)e^{2i\theta}. \qquad (12.3.5)$$

Since $z = x + iy$, this is

$$x = X(\tau) + R(\tau) \cos 2\theta, \qquad y = R(\tau) \sin 2\theta. \qquad (12.3.6)$$

Elimination of θ gives

$$(x - X(\tau))^2 + y^2 = (R(\tau))^2. \qquad (12.3.7)$$

The fluid attains a given speed q at points (x, y) on the circle (12.3.7) with $\tau = q^2$. A streamline corresponds to a given value of the stream function ψ (i.e., to a relation $\tau = a \sin^2 \theta$ for fixed a), by (12.3.1a). Therefore by (12.3.6) the streamlines are curves in the (x, y) plane given parametrically in terms of θ by

$$\begin{aligned}
x &= X(a \sin^2 \theta) + R(a \sin^2 \theta) \cos 2\theta, \\
y &= R(a \sin^2 \theta) \sin 2\theta.
\end{aligned} \qquad (12.3.8)$$

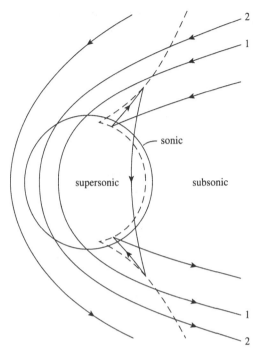

Fig. 12.3.1. Ringleb's flow, subsonic outside a circle and supersonic inside. On the limit line (– – –) the streamlines have cusps. The flow between streamlines 1 and 2, for example, could be the smooth flow of an ideal fluid in a channel.

Cusps occur where $dx/d\theta = dy/d\theta = 0$. The resulting streamline pattern is shown in Figure 12.3.1. Full details of the flow may be obtained in principle by calculating, in terms of (x, y), first (τ, θ) from (12.3.6), then (ϕ, ψ) from (12.3.1), and finally (u, v) from (12.1.3). The mapping between the physical plane (x, y) and the hodograph plane (u, v) has a fold in the physical plane, called a limit line, at which fluid apparently returns with a different velocity into a region from which it has come. Since fluid cannot have two velocities at a point, the mapping between (x, y) and (u, v) near a limit line cannot correspond to an actual flow, and taken as a whole the flow represented by Equations (12.3.6) and Figure 12.3.1 is impossible. We shall say more about limit lines in the next section. Here we note that in restricted regions the flow is not only possible but of interest. For example, deceleration of fluid through sonic speed usually produces a shock, but this can be avoided if the boundary of the fluid has a special shape. Such is provided by a channel with walls at streamlines 1 and 2 in Figure 12.3.1, assuming that boundary-layer effects may be neglected. The flow can then occur according to the formulae we have obtained, so that in the channel the speed of a fluid element increases from subsonic to supersonic and then decreases back to subsonic without the production of a shock. Thus Ringleb's flow captures some physically important properties of exact solutions of the full nonlinear equations of high speed flow of an ideal fluid.

12.4 Geometrical Theory. The Legendre Transformation

The hodograph transformation is an example of a general type of transformation for which there is a geometrical theory. Let variables u and v, functions of x and y, satisfy two simultaneous partial differential equations in which the coefficients are functions only of u and v. The equations, with coefficients $a_1, b_1, c_1, d_1, a_2, b_2, c_2$, and d_2, and with partial differentiation of u and v denoted by subscripts x and y, are taken to be

$$a_1 u_x + b_1 u_y + c_1 v_x + d_1 v_y = 0,$$
$$a_2 u_x + b_2 u_y + c_2 v_x + d_2 v_y = 0. \tag{12.4.1}$$

Because the coefficients depend on u and v, the equations are nonlinear. We shall write the differential relations $du = u_x dx + u_y dy$ and $dv = v_x dx + v_y dy$ in matrix form as

$$\begin{pmatrix} du \\ dv \end{pmatrix} = \begin{pmatrix} u_x & u_y \\ v_x & v_y \end{pmatrix} \begin{pmatrix} dx \\ dy \end{pmatrix}. \tag{12.4.2}$$

The determinant of the matrix in this equation is the Jacobian of the transformation between (x, y) and (u, v) and will be denoted j. Thus

$$j = u_x v_y - u_y v_x. \tag{12.4.3}$$

Consider a region in which $j \neq 0$. Then (12.4.2) has an inverse, namely

$$\begin{pmatrix} dx \\ dy \end{pmatrix} = \frac{1}{j} \begin{pmatrix} v_y & -u_y \\ -v_x & u_x \end{pmatrix} \begin{pmatrix} du \\ dv \end{pmatrix}. \tag{12.4.4}$$

Another way of obtaining (dx, dy) in terms of (du, dv) is to start with the inverse functions, that is, x and y as functions of u and v. Then with partial differentiation of x and y denoted by subscripts u and v, the differential relations $dx = x_u du + x_v dv$ and $dy = y_u du + y_v dv$ may be written

$$\begin{pmatrix} dx \\ dy \end{pmatrix} = \begin{pmatrix} x_u & x_v \\ y_u & y_v \end{pmatrix} \begin{pmatrix} du \\ dv \end{pmatrix}. \tag{12.4.5}$$

Since (12.4.4) and (12.4.5) refer to the same transformation, we must have

$$\begin{pmatrix} x_u & x_v \\ y_u & y_v \end{pmatrix} = \frac{1}{j} \begin{pmatrix} v_y & -u_y \\ -v_x & u_x \end{pmatrix}. \tag{12.4.6}$$

The four relations in (12.4.6) may be used to rewrite (12.4.1) in terms of x_u, x_v, y_u, and y_v. The result, after division by the Jacobian j, which we assumed to be nonzero, is

$$
\begin{aligned}
a_1 y_v - b_1 x_v - c_1 y_u + d_1 x_u &= 0, \\
a_2 y_v - b_2 x_v - c_2 y_u + d_1 x_u &= 0.
\end{aligned}
\tag{12.4.7}
$$

Because the coefficients depend only on u and v, these equations are linear. Thus we have shown that, when $j \neq 0$ in a region, each solution of the nonlinear equations (12.4.1) gives rise to a solution of the linear equations (12.4.7). In the reverse direction, let the Jacobian of the transformation between (u, v) and (x, y) be J. Thus

$$J = x_u y_v - x_v y_u. \tag{12.4.8}$$

By a basic property of Jacobians, we have $jJ = 1$ wherever j and J are nonzero. A similar argument to that just presented shows that, when $J \neq 0$ in a region, each solution of the linear equations (12.4.7) gives rise to a solution of the nonlinear equations (12.4.1).

In transforming between (12.4.1) and (12.4.7) in a region, much depends on whether j or J is ever zero. For example, if j is identically zero in a whole region of the (x, y) plane then (u, v) represents a simple wave as defined in Section 10.5. In general, the conditions $j = 0$ or $J = 0$ are satisfied on curves in the (x, y) and (u, v) planes and correspond either to multivalued solutions u, v or to the nonexistence of solutions u, v. The conditions $j = 0$ or $J = 0$ may also be satisfied at isolated points.

When $j = 0$ on a curve in the (x, y) plane, as in Figure 12.4.1a, we may imagine that the curve is the centreline of a strip in the (x, y) plane, which is then folded along this centreline and, after smooth stretching, is placed in the (u, v) plane. The curve in the (x, y) plane on which $j = 0$ is called a transition curve, and its image in the (u, v) plane is called an edge. Points (u, v) on one side of the edge are not images of points (x, y) close to the transition curve; this side is marked "void" in Figure 12.4.1b. The other side, marked "double," is covered doubly by the strip surrounding the transition curve in the (x, y) plane. Similarly, when $J = 0$ on a curve in the (u, v) plane, as in Figure 12.4.1d, we may imagine folding a strip in the (u, v) plane along this curve and placing it

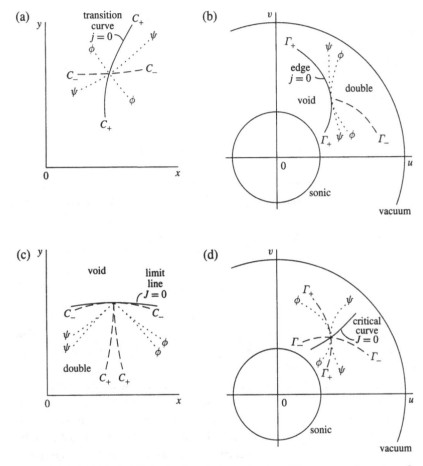

Fig. 12.4.1. (a), (b) A fold in the (x, y) plane at a transition curve, where $j = 0$, covers the (u, v) plane doubly on one side of an edge. (c), (d) A fold in the (u, v) plane at a critical curve, where $J = 0$, covers the (x, y) plane doubly on one side of a limit line. The uncovered sides of the edge and limit line are marked "void." The streamline ψ and potential line ϕ are orthogonal in (a), (c), and tangential in (b), (d); the characteristic C_+ in (a), (c) is orthogonal to the characteristic Γ_- in (b), (d), and similarly for C_- and Γ_+; the streamline and potential line in (a), (c) bisect the angles between $C_\pm$, and in (b), (d) they are tangential to a Γ characteristic. An edge is a Γ characteristic, a transition curve is a C characteristic, and a limit line is an envelope of C characteristics. The Γ characteristics, which depend on the critical speed $\hat{c}$ but not on the flow, are epicycloids filling the region between the sonic circle $q = \hat{c}$ and the circle $q = q_m = ((\gamma + 1)/(\gamma - 1))^{1/2}\hat{c}$ corresponding to the vacuum condition, that is, they fill the annulus $\hat{c}^2 < u^2 + v^2 < ((\gamma + 1)/(\gamma - 1))\hat{c}^2$.

in the (x, y) plane. The curve $J = 0$ in the (u, v) plane is a critical curve, and its image in the (x, y) plane is a limit line. The side of the limit line that is not the image of points in the (u, v) plane near the critical curve is marked "void" in Figure 12.4.1c. The other side, marked "double," is covered doubly by the strip surrounding the critical curve in the (u, v) plane.

Now assume that Equations (12.4.1) are those for steady flow of a polytropic gas, as given in Sections 12.1 and 12.2. The Mach-line characteristics in the (x, y) plane of the partial differential equations for the velocity components $u(x, y)$ and $v(x, y)$ will be called C characteristics, and the two families will be denoted C_+ and C_-. Similarly, the Mach-line characteristics in the (u, v) plane of the equations for $x(u, v)$ and $y(u, v)$ will be called Γ characteristics, and the two families will be denoted Γ_+ and Γ_-. The C and Γ characteristics correspond to each other under the hodograph transformation; that is, the image of a C_+ characteristic is a Γ_+ characteristic, and the image of a C_- characteristic is a Γ_- characteristic. The direction of a C_+ characteristic at a point (x, y) is perpendicular to the direction of the Γ_- characteristic at the corresponding point (u, v), and similarly for C_- and Γ_+. It follows that the velocity component parallel to a Γ characteristic is c, since this is the velocity component perpendicular to a C characteristic. The occurrence and location of C characteristics depend on the solution $u(x, y)$ and $v(x, y)$ being considered, because of the nonlinearity of the differential equations for u and v, but the Γ characteristics are fixed, because of the linearity of the equations for $x(u, v)$ and $y(u, v)$, and depend only on the Bernoulli constant in (12.2.1), for example on the critical speed $\hat{c}$ or the maximum speed $q_m = ((\gamma + 1)/(\gamma - 1))^{1/2}\hat{c}$. Since $q = (u^2 + v^2)^{1/2}$, these two speeds determine two circles centred on the origin in the (u, v) plane, the sonic circle $q = \hat{c}$ and the vacuum-condition circle $q = q_m$, as shown in Figures 12.4.1b and d. We shall later give the ordinary differential equation for the Γ_+ and Γ_- characteristics. The equation may be solved exactly, and the solution shows that these characteristics are the epicycloids traced out by a point on the circumference of a circle of diameter $q_m - \hat{c}$ that rolls without slipping on the sonic circle $q = \hat{c}$. Thus the Γ characteristics doubly cover the annulus $\hat{c} < q < q_m$ between the sonic circle and the vacuum-condition circle. If the arcs of Γ_+ and Γ_- in Figures 12.4.1b and d are continued as far as possible, they meet the sonic circle at right angles and the vacuum-condition circle tangentially. The complete set of characteristics is obtained by giving these two extended arcs every possible rotation about the origin of the (u, v) plane.

A transition curve is always a C characteristic, at which the flow is therefore sonic or supersonic, and its image in the (u, v) plane, the corresponding edge, is therefore a Γ characteristic. The C characteristics of the other family map doubly to the Γ characteristics of the other family, which meet the edge nontangentially and on one side only. Streamlines and potential lines in the (x, y) plane intersect at right angles, bisecting the angles between the C characteristics. The corresponding streamlines and potential lines in the (u, v) plane are tangential to each other and to the edge, as shown in Figure 12.4.1b. A transition curve can occur in a real flow, and occurs for example in high speed flow in a nozzle, because no impossibility arises when the same velocity is attained by fluid at different points.

A limit line is tangential to the C characteristics of one family, for which it forms an envelope, and so can exist only where the flow is sonic or supersonic. It is a locus of cusps on the C characteristics of the other family. The streamlines (i.e., lines of constant ψ) and potential lines (i.e., lines of constant ϕ) also have cusps on the limit line, and they bisect the angles between the characteristics, as shown in Figure 12.4.1c. In the (u, v) plane, the corresponding streamlines and potential lines are tangential to each other and to one family of Γ characteristics at the critical curve, as shown in Figure 12.4.1d. A limit line cannot occur in a real flow, because the folding of the (u, v) plane assigns two velocities to points (x, y) on one side of the line and no velocity to points on the other side. The flow may then contain a shock, but its position, which is not that of the limit line, cannot be determined by an extension of the hodograph method and requires a fresh analysis of the governing equations. Thus the occurrence of limit lines restricts the scope of the hodograph method in the solution of realistic problems in high speed flow.

Many of the statements just made have been proved already, in Chapter 10. When the general equations (12.4.1) are the fluid-dynamical equations (10.1.4) for steady, two-dimensional, irrotational, homentropic flow, the equations (12.4.7) for $x(u, v)$ and $y(u, v)$ are

$$x_v - y_u = 0, \tag{12.4.9a}$$

$$(c^2 - v^2)x_u + uv(x_v + y_u) + (c^2 - u^2)y_v = 0. \tag{12.4.9b}$$

Here c^2 is a function of $q^2 = u^2 + v^2$, by Bernoulli's equation (12.2.1). The differential equation for the characteristics of (12.4.9) (i.e., the Γ characteristics) may be found by the method of Section 10.2. It is (10.3.5):

$$c^2(du^2 + dv^2) = (u\,du + v\,dv)^2. \tag{12.4.10}$$

This may be compared with Equation (10.2.8) for the C characteristics:

$$c^2(dx^2 + dy^2) = (v\,dx - u\,dy)^2. \tag{12.4.11}$$

The orthogonality property of (C_+, Γ_-) and of (C_-, Γ_+) is expressed by (10.3.7), that is, $(dy/dx)_+(dv/du)_- = -1$ and $(dy/dx)_-(dv/du)_+ = -1$, as illustrated in Figure 10.3.1. Equation (12.4.10) for the Γ characteristics may be solved by the method given in Section 10.3 for the C characteristics, that is, by expressing the velocity field in polar coordinates (q, θ) as $(u, v) = (q \cos \theta, q \sin \theta)$ and using angles wherever possible. The solution gives the epicycloids forming the Γ characteristics.

Equations (12.4.9) may be used to express $J = x_u y_v - x_v y_u$ as a quadratic form in (x_u, x_v) or (y_u, y_v). For example, substituting for y_u from (12.4.9a) and y_v from (12.4.9b) we obtain

$$
\begin{aligned}
J &= \frac{(c^2 - v^2)x_u^2 + 2uv x_u x_v + (c^2 - u^2)x_v^2}{-(c^2 - u^2)} \\
&= \frac{\{(c^2 - v^2)x_u + uv x_v\}^2 + c^2(c^2 - u^2 - v^2)x_v^2}{-(c^2 - u^2)(c^2 - v^2)}.
\end{aligned}
\tag{12.4.12}
$$

Therefore $J < 0$ when $u^2 + v^2 < c^2$, which establishes our earlier remark that a limit line cannot occur in subsonic flow. Similarly, Equations (10.1.4) give $j = u_x v_y - u_y v_x$ as a quadratic form in (u_x, u_y) or (v_x, v_y), for example

$$
\begin{aligned}
j &= \frac{(c^2 - u^2)u_x^2 - 2uv u_x u_y + (c^2 - v^2)u_y^2}{-(c^2 - v^2)} \\
&= \frac{\{(c^2 - u^2)u_x - uv u_y\}^2 + c^2(c^2 - u^2 - v^2)u_y^2}{-(c^2 - u^2)(c^2 - v^2)}.
\end{aligned}
\tag{12.4.13}
$$

Therefore $j < 0$ when $u^2 + v^2 < c^2$, unless $u_x = u_y = 0$, so that a transition curve cannot occur in a nonconstant subsonic flow.

The relative orientations of the curves in Figure 12.4.1 may be established by determining the leading order terms in Taylor-series expansions, or in fractional-power expansions, of the quantities $x(u, v)$, $y(u, v)$, $u(x, y)$, and $v(x, y)$ near the lines on which $j = 0$ or $J = 0$. For details we refer the reader to the bibliographic notes.

The use of u and v as independent variables suggests the definition of further quantities. By analogy with the potential $\phi(x, y)$ and stream function $\psi(x, y)$, defined so that their partial derivatives with respect to x and y are simply related to (u, v) and $(\rho u, \rho v)$, it is natural to define a potential-like function $\Phi(u, v)$ and a stream-like function $\Psi(\rho u, \rho v)$ whose partial derivatives with respect to u and v, or ρu and ρv, are simply related to (x, y). These functions are so chosen that their partial derivatives Φ_u, Φ_v, $\Psi_{\rho u}$, and $\Psi_{\rho v}$

satisfy

$$(x, y) = (\Phi_u, \Phi_v) = (-\Psi_{\rho v}, \Psi_{\rho u}). \qquad (12.4.14)$$

The relations between (x_u, x_v, y_u, y_v) and (u_x, u_y, v_x, v_y) given in (12.4.6) show that

$$\Phi = ux + vy - \phi, \qquad \Psi = \rho uy - \rho vx - \psi. \qquad (12.4.15)$$

These equations express the fact that Φ is obtained from ϕ by a Legendre transformation, and vice versa, and similarly for Ψ and ψ. The hodograph transformation between (x, y) and (u, v) corresponds to these Legendre transformations. Therefore directly, or by the theory of the Legendre transformation, we may obtain numerous relations among the variables (x, y), (u, v), (q, θ), (ϕ, ψ), (Φ, Ψ), and (j, J) and apply them to problems in high speed flow.

12.5 Bibliographic Notes

Early papers on the hodograph method in Table 13.2.1 are Molenbroek (1890) and Chaplygin (1904). A later set of four papers on the method is Lighthill (1947). Reference articles in Table 13.3.1 on the method are Lighthill (1953a), Kuo & Sears (1955), and Schiffer (1960). Texts and monographs in Table 13.4.1 with full accounts of the method are Courant & Friedrichs (1948), Shapiro (1953, 1954), von Mises (1958), Curle & Davies (1971), Manwell (1971), and Landau & Lifshitz (1987); see also Milne-Thomson (1968). For the Legendre transformation, see Courant & Hilbert (1962). The trend towards computational fluid dynamics has rather eclipsed the hodograph method, but for a recent example of its use see Ardalan, Meiron & Pullin (1995).

12.6 Further Results and Exercises

1. Write down the hodograph equations, in standard notation (τ, θ) and (ϕ, ψ), for the motion of a polytropic gas with ratio of specific heats γ. Find all the solutions in which the stream function ψ is proportional to $\sin \theta$, and show that, for one of them,

$$\psi \propto \tau^{-\frac{1}{2}} (1 - \tau)^{\beta+1} \sin \theta.$$

(This is Temple's solution.) Determine the corresponding potential ϕ. Analyse the flow as far as you can, particularly with regard to the overall shape of the streamlines and the occurrence of any limit lines.

2. Show that the equations of motion for steady irrotational fluid flow may be written, in standard notation (x, y), (q, θ), (ϕ, ψ), and (ρ, ρ_0), as

$$x_\phi = \frac{\cos\theta}{q}, \qquad x_\psi = -\frac{\rho_0 \sin\theta}{\rho q}, \qquad y_\phi = \frac{\sin\theta}{q}, \qquad y_\psi = \frac{\rho_0 \cos\theta}{\rho q}.$$

Let the speed at which the flow would become sonic be $\hat{c}$, and define ζ and $F(\zeta)$ by

$$\zeta = \frac{1}{\rho_0} \int_{\hat{c}}^{q} \rho \frac{dq}{q}, \qquad F(\zeta) = -q^2 \frac{\rho_0}{\rho} \frac{d}{dq}\left(\frac{\rho_0}{\rho q}\right).$$

Show that, provided the Jacobian of the transformation between (x, y) and (ϕ, ψ) is neither zero nor infinite, the equations of motion may be reduced to $\psi_{\zeta\zeta} + F(\zeta)\psi_{\theta\theta} = 0$. For constants k, n, and λ, and for an arbitrary linear combination $C_{1/(2n+3)}$ of Bessel functions $J_{1/(2n+3)}$ and $Y_{1/(2n+3)}$, show that if $F(\zeta) = -k^2\zeta^{2n+1}$ then the equations of motion have a solution

$$\psi(\zeta, \theta) = \zeta^{\frac{1}{2}} C_{1/(2n+3)}\left(\frac{2\lambda k}{2n+3}\zeta^{n+\frac{3}{2}}\right) e^{-i\lambda\theta}.$$

3. (See Courant & Friedrichs 1948.) In the theory of the hodograph transformation, prove as many results as you can about the relations between streamlines, potential lines, limit lines, critical curves, and transition curves. Where appropriate, relate your results to envelopes of curves. How are your results related to characteristics in the physical plane?

4. (See von Mises 1958.) Use the hodograph transformation to analyse the following two-dimensional compressible flows: point source, point sink, vortex flow, and spiral flow. Give particular attention to the occurrence of limit lines and to regions of space in which your formulae do not correspond to a flow that could occur in practice.

5. For high speed flow past an aerofoil, choose a few simple curves in the physical plane, for example straight lines parallel to the unperturbed flow, and similarly straight lines perpendicular to it, and determine the images of these curves in the hodograph plane. Sketch also in the hodograph plane the whole image of the physical plane.

6. Derive from first principles the hodograph equations for a polytropic gas, and show that, in notation such that Bernoulli's equation is $\rho = \rho_0(1 - \tau)^\beta$, they are

$$2\tau(1 - \tau)^{\beta+1}\phi_\tau = -(1 - (2\beta + 1)\tau)\psi_\theta, \qquad (1 - \tau)^\beta\phi_\theta = 2\tau\psi_\tau. \qquad (12.6.1)$$

What choice of ϕ and ψ is suitable for describing the flow near a point source in two dimensions? Analyse the flow as far as you can.

13

Guide to High Speed Flow

13.1 Introduction

This chapter is a guide to published work on high speed flow, classified into (a) early research, (b) reference works, (c) texts and monographs, (d) surveys and reviews, and (e) recent research. In each category, the years covered are restricted to a period of particular interest. The work is arranged in tables in which the entries are listed by year and author. Full details of each entry are given in the references, listed alphabetically by author, at the end of the book. The tables complement the bibliographic notes, arranged by topic, at the end of the chapters. Although the references are selective rather than complete, they include a high proportion of the more important works on high speed flow and provide ready access to the entire literature of the subject.

13.2 Research on High Speed Flow, 1860–1945

Table 13.2.1 lists some contributions to research in high speed flow from the middle of the nineteenth century to the end of the Second World War. For a brief account of research in this period, see Sections 1.4 and 1.6.

13.3 Reference Works on High Speed Flow, 1953–1964

Table 13.3.1 lists a selection of articles from the three reference works (a) *Modern Developments in Fluid Dynamics: High Speed Flow*, Volumes 1, 2 (ed. L. Howarth, 1953); (b) *High Speed Aerodynamics and Jet Propulsion* ("the Princeton Series," 12 volumes, 1955–1964); and (c) the *Encyclopedia of Physics* (ed. S. Flügge, volumes 8–9, 1959–1960). Further information about these works is given in Section 1.6 and on the first page of the references.

Table 13.2.1. *Research on high speed flow, 1860–1945*

1860	S. Earnshaw	1931	A. Busemann
1860	B. Riemann	1933	G. I. Taylor & J. W. Maccoll
1870	W. J. M. Rankine	1934	W. F. Durand
1878	E. Mach	1936	L. Prandtl
1887	E. Mach & P. Salcher	1937	L. Crocco
1889	H. Hugoniot	1939	H. S. Tsien
1890	P. Molenbroek	1940	F. von Ringleb
1893	W. Sutherland	1941	Th. von Kármán
1904	S. A. Chaplygin	1941	H. G. Küssner
1908	Th. Meyer	1942	H. A. Bethe
1910	Lord Rayleigh	1942	K. G. Guderley
1910	G. I. Taylor	1942	K. Oswatitsch
1913	O. Janzen	1943	J. von Neumann
1916	Lord Rayleigh	1944	L. D. Landau
1927	J. Ackeret	1944	M. J. Lighthill
1928	H. Glauert	1945	F. I. Frankl

Table 13.3.1. *Reference works on high speed flow, 1953–1964*

1953	1955	1958	1960
W. G. Bickley	A. Ferri a,b	L. Crocco	H. Cabannes
W. F. Cope	K. O. Friedrichs	W. D. Hayes	R. E. Meyer
L. Howarth	M. A. Heaslet &	A. R. Kantrowitz	M. Schiffer
C. R. Illingworth	H. Lomax	H. Polachek &	R. Timman
G. J. Kynch	Th. von Kármán	R. J. Seeger	
M. J. Lighthill a	Y. H. Kuo &	H. G. Stever	**1961**
W. A. Mair &	W. R. Sears	H. S. Tsien	
J. A. Beavan	M. J. Lighthill		W. Bleakney &
R. E. Meyer	F. D. Rossini	**1959**	R. J. Emrich
O. A. Saunders	W. R. Sears		F. E. Goddard
H. B. Squire		K. Oswatitsch	J. E. Smith
G. Temple	**1957**	J. Serrin	
G. N. Ward			**1964**
A. D. Young	J. C. Evvard		
	C. W. Frick		F. K. Moore
	I. E. Garrick		
	R. T. Jones &		
	D. Cohen		

13.4 Texts and Monographs on High Speed Flow, 1947–1998

Table 13.4.1 lists a selection of the better-known texts and monographs on high speed flow published since the end of the Second World War. The table includes a few texts not wholly concerned with high speed flow.

Table 13.4.1. *Texts and monographs on high speed flow, 1947–1998*

1947	H. W. Liepmann & A. E. Puckett	1966	Ya. B. Zel'dovich & Yu. P. Raizer
1947	R. Sauer	1967	Ya. B. Zel'dovich & Yu. P. Raizer
1948	R. Courant & K. O. Friedrichs	1968	C. Ferrari & F. G. Tricomi
1949	A. Ferri	1971	S. N. Curle & H. J. Davies
1953	A. H. Shapiro	1971	A. R. Manwell
1954	A. H. Shapiro	1974	G. B. Whitham
1955	G. N. Ward	1975	M. Van Dyke
1956	K. Oswatitsch	1976	M. E. Goldstein
1957	H. W. Liepmann & A. Roshko	1982	J. D. Anderson
1958	L. Bers	1983	H. Ockendon & A. B. Tayler
1958	R. von Mises	1985	P. A. Thompson
1959	J. W. Miles	1986	J. D. Cole & L. P. Cook
1959	S.-I. Pai	1987	L. D. Landau & E. M. Lifshitz
1961	M. T. Landahl	1989	J. D. Anderson
1962	K. G. Guderley	1992	G. Ben-Dor
1964	K. Stewartson	1993	P. Prasad
1965	W. G. Vincenti & C. H. Kruger	1994	I. I. Glass & J. P. Sislian
1966	W. D. Hayes & R. F. Probstein	1998	C. B. Laney

Table 13.5.1. *Surveys and reviews of topics in high speed flow, 1949–1999*

1949	M. J. Lighthill	1986	H. Hornung
1952	P. Germain & R. Bader	1986	P. L. Roe
1953	G. Guderley	1986	J. T. Stuart
1956	M. J. Lighthill	1987	S. S. Kutateladze,
1964	P. Germain		V. E. Nakoryakov &
1964	I. Teipel		A. A. Borisov
1969	Ya. B. Zel'dovich & Yu. P. Raizer	1987	G. Moretti
1970	J. W. Rich & C. E. Treanor	1988	G. Ben-Dor
1971	A. Busemann	1989	R. Menikoff & B. J. Plohr
1971	W. D. Hayes	1993	H. K. Cheng
1971	V. V. Mikhailov, V. Ya. Neiland &	1993	G. A. Tirsky
	V. V. Sychev	1994	S. K. Lele
1973	G. Y. Nieuwland & B. M. Spee	1994	E. F. Spina, A. J. Smits &
1976	V. V. Rusanov		S. K. Robinson
1977	P. Bradshaw	1995	E. J. Gutmark, K. C. Schadow &
1977	J. E. Ffowcs Williams		K. H. Yu
1978	O. S. Ryzhov	1995	C. K. W. Tam
1979	R. W. MacCormack & H. Lomax	1996	E. T. Curran, W. H. Heiser &
1980	T. C. Adamson Jr. & A. F. Messiter		D. T. Pratt
1980	H. Tijdeman & R. Seebass	1996	C. Ferrari
1981	W. C. Griffith	1997	V. L. Wells & R. A. Renaut
1981	V. A. Solonnikov & A. V. Kazhikhov	1998	J. P. Bonnet, D. Grésillon &
1981	D. A. Sullivan		J. P. Taran
1982	D. A. Caughey	1998	M. S. Ivanov & S. F. Gimelshein
1983	M. B. Lesser & J. E. Field	1999	R. Agarwal
1983	H. Reichenbach	1999	P. A. Gnoffo
1984	H. Sobieczky & A. R. Seebass	1999	E. J. Gutmark & F. F. Grinstein
1985	N. Rott	1999	E. Turkel

13.5 Surveys and Reviews of High Speed Flow, 1949–1999

Table 13.5.1 lists a selection of surveys and reviews of specialised topics in high speed flow, published since 1949. Most were published in the *Annual Review of Fluid Mechanics*.

13.6 Research on High Speed Flow, 1990–1998

Tables 13.6.1a–d list all the papers on high speed flow published in the *Journal of Fluid Mechanics* in the period 1990–1998, classified into the topics (a) shock waves; (b) hypersonic flow; (c) jets, boundary layers, shear layers, and mixing

Table 13.6.1a. *Research on high speed flow, 1990–1998: Shock waves. The first author of each paper is listed*

1990	1994	1997
P. Colella	M. Brouillette	J. E. Broadwell
M. S. Cramer	S. C. Gülen	B. P. Brown
J. W. Grove	M. Mond	G. Erlebacher
M. Wang	R. Samtaney	L. F. Henderson
	A. Sasoh	G. Jourdan
1991	M. Watanabe	N. K.-R. Kevlahan
	J. Yang	S. Lee
E. M. Barkhudarov		H. Li (2)
M. S. Cramer	1995	K. Mahesh
L. F. Henderson		S. G. Mallinson
D. Klages	A. Chpoun	J. Olejniczak
T. J. McIntyre	C. F. Delale	A. G. Panaras
G. C. Pham-Van-Diep	R. L. Holmes	
J. B. Young	J. W. Jacobs	1998
	K. Mahesh	
1992	J. Vuillon	R. F. Chisnell
		L. F. Henderson
N. K. Bourne	1996	G. E. Reisman
A. Guha		N. J. Zabusky
J. W. Jacobs	N. Apazidis	
R. J. Stalker	D. R. Buttsworth	
	S. C. Crow	
1993	Z. Ding	
	M. D. Fox	
I. D. Boyd	O. Igra	
P. W. Hammerton	N. K.-R. Kevlahan	
S. Lee	A. Korobkin	
B. W. Skews	A. Levy	
M. Spiegelman	J. J. Liu	
	J. J. Quirk	
	B. W. Skews	

Table 13.6.1b. *Research on high speed flow, 1990–1998:*
Hypersonic flow. The first author of each paper is listed

1990	1994	1996
S. N. Brown	Y. Fu	K. W. Cassel
S. J. Cowley	W. C. Griffith	M. A. Gallis
M. E. Goldstein	R. M. Kerimbekov	
M. N. MacCrossan	A. F. Khorrami	1997
A. F. Messiter	M. N. MacCrossan	
P. Riesco-Chueca		S. O. Seddougui
	1995	S. Séror
1993		
	D. R. Buttsworth	
N. D. Blackaby	G. Simeonides	
Y. Fu		
S. E. Grubin (2)		

Table 13.6.1c. *Research on high speed flow, 1990–1998: Jets, boundary*
layers, shear layers, and mixing layers. The first author of each paper is listed

1990	1992	1995	1997
T. F. Balsa	P. Balakumar (2)	G. M. Bassett	F. Bastin
P. W. Duck (3)	J. Bridges	K. W. Cassel	T. Colonius
E. M. Fernando	J. M. Hyun	N. T. Clemens	J. W. Naughton
Y. P. Guo	M. Wang	A. Dillmann	G. Raman (2)
T. L. Jackson		J. He	O. S. Ryzhov
A. D. Kosinov	1993	S. J. Leib	A. Simone
D. M. Moody		S. Sarkar	D. R. Smith
P. J. Morris	G. A. Blaisdell		
F. T. Smith	H. G. Im	1996	1998
C. K. W. Tam	G. Lu		
	C. K. W. Tam	N. A. Adams	S. A. Arnette
1991		P. Ghosh Choudhuri	F. P. Bertolotti
	1994	P. W. Hammerton	O. M. Haddad
C. E. Grosch		H. G. Im	W. Konrad
T. L. Jackson	S. Barre	A. Korobkin	M. F. Miller
S. J. Leib	I. D. Boyd	C.-Y. Kuo	J. Panda
M. R. Malik	L. Brusniak	G. Raman	G. Raman
J. A. Pedelty	C.-L. Chang	H. S. Ribner	C. K. W. Tam
N. D. Sandham	J. F. Donovan	P. J. Strykowski	
I. G. Shukhman	P. W. Duck	A. W. Vreman	
E. F. Spina	I. P. Vickers	M. Wang	
C. K. W. Tam			

Table 13.6.1d. *Research on high speed flow, 1990–1998: Miscellaneous topics, including fans, cascades, propellers, aerofoils, plates, blunt bodies, ducts, nozzles, transonic flow, two-phase flow, vortices, and rapid distortion theory. The first author of each paper is listed*

1990	1993	1995	1997
S. D. Heister	R. K. Amiet	R. K. Amiet	A. T. Fedorchenko
R. Ishii	H. R. van den Berg	K. Ardalan	M. S. Howe
	G. Buresti	N. Botta	F. G. Leppington
1991	C. Cambon	G. N. Coleman	J. F. Monaco
	A. Kluwick	P. G. Huang	M. R. Myers
H. Ardavan	N. Peake	B. E. Mitchell	N. Peake (2)
T. Colonius	M. C. A. M. Peters	M. R. Myers	Z. Rusak
N. Peake	Z. Rusak	A. B. Parry	
J. R. Prakash	M. Tanaka	A. Paull	**1998**
S. Sarkar	J.-Z. Wu	D. Virk	
W. C. Selerowicz			C. F. Delale
F. T. Smith	**1994**	**1996**	I. M. Kalkhoran
K. H. Yu			E. C. Magi
	H. Ardavan	C.-C. Chang	S. J. Majumdar
1992	A. Dillmann	E. A. Cox	F. Mashayek
	A. Korobkin	A. Goldshtein	D. W. Moore
M. S. Cramer	N. Peake (2)	D. D. Joseph	C. K. W. Tam
P. A. Durbin	J. A. K. Stott	A. Kluwick	
G. Erlebacher		L. Likhterov	
A. Korobkin		A. Shajii	
A. G. Panaras		M. Tomasini	
N. Peake (2)			

layers; and (d) miscellaneous topics, including fans, cascades, propellers, aero-foils, plates, blunt bodies, ducts, nozzles, transonic flow, two-phase flow, vortices, and rapid distortion theory. The first author of each paper is given in the tables, and the remaining authors, with full details of the papers, are given in the references. Each paper appears only once in the tables. For example, papers on hypersonic boundary layers appear in Table 13.6.1b, on hypersonic flow. Some topics, for example aeroacoustics and turbulence, occur in papers distributed throughout the four tables.

References

The Princeton Series, the *Encyclopedia of Physics*, and *Modern Developments*

The twelve volumes of *High Speed Aerodynamics and Jet Propulsion* (Oxford University Press; © Princeton University Press), with their editors and year of publication, are:

1. *Thermodynamics and Physics of Matter*. F. D. Rossini. 1955.
2. *Combustion Processes*. B. Lewis, R. N. Pease, and H. S. Taylor. 1956.
3. *Fundamentals of Gas Dynamics*. H. W. Emmons. 1958.
4. *Theory of Laminar Flows*. F. K. Moore. 1964.
5. *Turbulent Flows and Heat Transfer*. C. C. Lin. 1956.
6. *General Theory of High Speed Aerodynamics*. W. R. Sears. 1955.
7. *Aerodynamic Components of Aircraft at High Speeds*. A. F. Donovan and H. R. Lawrence. 1957.
8. *High Speed Problems of Aircraft and Experimental Methods*. A. F. Donovan, H. R. Lawrence, F. E. Goddard, and R. R. Gilruth. 1961.
9. *Physical Measurements in Gas Dynamics and Combustion*. R. W. Ladenburg, B. Lewis, R. N. Pease, and H. S. Taylor. 1955.
10. *Aerodynamics of Turbines and Compressors*. W. R. Hawthorne. 1964.
11. *Design and Performance of Gas Turbine Power Plants*. W. R. Hawthorne and W. T. Olson. 1960.
12. *Jet Propulsion Engines*. O. E. Lancaster. 1959.

Two volumes of the *Encyclopedia of Physics* (Springer-Verlag; ed. S. Flügge), and their coeditor, are:

Fluid Dynamics I. C. Truesdell. 1959.
Fluid Dynamics III. C. Truesdell. 1960.

Two volumes of *Modern Developments in Fluid Dynamics* (Oxford University Press), and their editor, are:

High Speed Flow, Volumes 1, 2. L. Howarth. 1953.

229

The above volumes will be referred to as the Princeton Series, the Encyclopedia of Physics, and Modern Developments. A typical reference to an individual article is Heaslet, M. A. & Lomax, H. 1955. Supersonic and transonic small perturbation theory. *Princeton Series* **6**, 122–344.

Ackeret, J. 1927. Gasdynamik. *Handbuch der Physik* VII, **5**, 289–342. Springer-Verlag.
Adams, N. A. & Kleiser, L. 1996. Subharmonic transition to turbulence in a flat-plate boundary layer at Mach number 4.5. *J. Fluid Mech.* **317**, 301–335.
Adamson, T. C., Jr. & Messiter, A. F. 1980. Analysis of two-dimensional interactions between shock waves and boundary layers. *Annu. Rev. Fluid Mech.* **12**, 103–138.
Agarwal, R. 1999. Computational fluid dynamics of whole-body aircraft. *Annu. Rev. Fluid Mech.* **31**, 125–169.
Amiet, R. K. 1993. On the second-order solution to the Sears problem for compressible flow. *J. Fluid Mech.* **254**, 213–228.
Amiet, R. K. 1995. Airfoil leading-edge suction and energy conservation for compressible flow. *J. Fluid Mech.* **289**, 227–242.
Anderson, J. D. 1982. *Modern Compressible Flow with Historical Perspective.* McGraw-Hill.
Anderson J. D. 1989. *Hypersonic and High Temperature Gas Dynamics.* McGraw-Hill.
Apazidis, N. & Lesser, M. B. 1996. On generation and convergence of polygonal-shaped shock waves. *J. Fluid Mech.* **309**, 301–319.
Ardalan, K., Meiron, D. I. & Pullin, D. I. 1995. Steady compressible vortex flows: The hollow-core vortex array. *J. Fluid Mech.* **301**, 1–17.
Ardavan, H. 1991. The breakdown of the linearized theory and the role of quadrupole sources in transonic rotor acoustics. *J. Fluid Mech.* **226**, 591–624.
Ardavan, H. 1994. Asymptotic analysis of the radiation by volume sources in supersonic rotor acoustics. *J. Fluid Mech.* **266**, 33–68.
Arnette, S. A., Samimy, M. & Elliott, G. S. 1998. The effects of expansion on the turbulence structure of compressible boundary layers. *J. Fluid Mech.* **367**, 67–105.
Balakumar, P. & Malik, M. R. 1992a. Discrete modes and continuous spectra in supersonic boundary layers. *J. Fluid Mech.* **239**, 631–656.
Balakumar, P. & Malik, M. R. 1992b. Waves produced from a harmonic point source in a supersonic boundary-layer flow. *J. Fluid Mech.* **245**, 229–247.
Balsa, T. F. & Goldstein, M. E. 1990. On the instabilities of supersonic mixing layers: A high-Mach-number asymptotic theory. *J. Fluid Mech.* **216**, 585–611.
Barkhudarov, E. M., Mdivnishvili, M. O., Sokolov, I. V., Taktakishvili, M. I. & Terekhin, V. E. 1991. Mach reflection of a ring shock wave from the axis of symmetry. *J. Fluid Mech.* **226**, 497–509.
Barre, S., Quine, C. & Dussauge, J. P. 1994. Compressibility effects on the structure of supersonic mixing layers: Experimental results. *J. Fluid Mech.* **259**, 47–78.
Bassett, G. M. & Woodward, P. R. 1995. Numerical simulation of nonlinear kink instabilities on supersonic shear layers. *J. Fluid Mech.* **284**, 323–340.
Bastin, F., Lafon, P. & Candel, S. 1997. Computation of jet mixing noise due to coherent structures: The plane jet case. *J. Fluid Mech.* **335**, 261–304.
Batchelor, G. K. 1967. *An Introduction to Fluid Dynamics.* Cambridge University Press.
Bauer, F., Garabedian, P. & Korn, D. 1972. *A Theory of Supercritical Wing Sections, with Computer Program and Examples.* Lecture Notes in Economics and Mathematical Systems **66**. Springer-Verlag.

Ben-Dor, G. 1987. A reconsideration of the three-shock theory for a pseudo-steady Mach reflection. *J. Fluid Mech.* **181**, 467–484.

Ben-Dor, G. 1988. Steady, pseudo-steady and unsteady shock wave reflections. *Progress in Aerospace Sciences* **25**, 329–412.

Ben-Dor, G. 1992. *Shock Wave Reflection Phenomena.* Springer-Verlag.

Ben-Dor, G. & Glass, I. I. 1979. Domains and boundaries of non-stationary oblique shock-wave reflexions. 1. Diatomic gas. *J. Fluid Mech.* **92**, 459–496.

Ben-Dor, G. & Glass, I. I. 1980. Domains and boundaries of non-stationary oblique shock-wave reflections. 2. Monatomic gas. *J. Fluid Mech.* **96**, 735–756.

Ben-Dor, G. & Takayama, K. 1985. Analytical prediction of the transition from Mach to regular reflection over cylindrical concave wedges. *J. Fluid Mech.* **158**, 365–380.

Bers, L. 1958. *Mathematical Aspects of Subsonic and Transonic Gas Dynamics.* Surveys in Applied Mathematics, volume 3. Wiley.

Bertolotti, F. P. 1998. The influence of rotational and vibrational energy relaxation on boundary-layer stability. *J. Fluid Mech.* **372**, 93–118.

Bethe, H. A. 1942. On the theory of shock waves for an arbitrary equation of state. Office Sci. Res. & Dev. Rep. 545, serial number NDRC-B-237, United States.

Bickley, W. G. 1953. Some exact solutions of the equations of steady homentropic flow of an inviscid gas. *Modern Developments* **1**, 158–189.

Blackaby, N. D., Cowley, S. J. & Hall, P. 1993. On the instability of hypersonic flow past a flat plate. *J. Fluid Mech.* **247**, 369–416.

Blaisdell, G. A., Mansour, N. N. & Reynolds, W. C. 1993. Compressibility effects on the growth and structure of homogeneous turbulent shear flow. *J. Fluid Mech.* **256**, 443–485.

Bleakney, W. & Emrich, R. J. 1961. The shock tube. *Princeton Series* **8**, 596–647.

Bonnet, J. P., Grésillon, D. & Taran, J. P. 1998. Nonintrusive measurements for high-speed, supersonic, and hypersonic flows. *Annu. Rev. Fluid Mech.* **30**, 231–273.

Botta, N. 1995. The inviscid transonic flow about a cylinder. *J. Fluid Mech.* **301**, 225–250.

Bourne, N. K. & Field, J. E. 1992. Shock-induced collapse of single cavities in liquids. *J. Fluid Mech.* **244**, 225–240.

Boyd, I. D. 1993. Temperature dependence of rotational relaxation in shock waves of nitrogen. *J. Fluid Mech.* **246**, 343–360.

Boyd, I. D., Beattie, D. R. & Cappelli, M. A. 1994. Numerical and experimental investigations of low-density supersonic jets of hydrogen. *J. Fluid Mech.* **280**, 41–67.

Bradshaw, P. 1977. Compressible turbulent shear layers. *Annu. Rev. Fluid Mech.* **9**, 33–54.

Bridges, J. & Hussain, F. 1992. Direct evaluation of aeroacoustic theory in a jet. *J. Fluid Mech.* **240**, 469–501.

Broadwell, J. E. 1997. Shocks and energy dissipation in inviscid fluids: A question posed by Lord Rayleigh. *J. Fluid Mech.* **347**, 375–380.

Brouillette, M. & Sturtevant, B. 1994. Experiments on the Richtmyer–Meshkov instability: Single-scale perturbations on a continuous interface. *J. Fluid Mech.* **263**, 271–292.

Brown, B. P. & Argrow, B. M. 1997. Two-dimensional shock tube flow for dense gases. *J. Fluid Mech.* **349**, 95–115.

References

Brown, S. N., Cheng, H. K. & Lee, C. J. 1990. Inviscid–viscous interaction on triple-deck scales in a hypersonic flow with strong wall cooling. *J. Fluid Mech.* **220**, 309–337.

Brusniak, L. & Dolling, D. S. 1994. Physics of unsteady blunt-fin-induced shock wave/turbulent boundary layer interactions. *J. Fluid Mech.* **273**, 375–409.

Buresti, G. & Casarosa, C. 1993. The one-dimensional adiabatic flow of equilibrium gas–particle mixtures in variable-area ducts with friction. *J. Fluid Mech.* **256**, 215–242.

Busemann, A. 1931. Gasdynamik. *Handbuch der Experimentalphysik* **4**, 341–460. Akad. Verlag., Leipzig.

Busemann, A. 1971. Compressible flow in the thirties. *Annu. Rev. Fluid Mech.* **3**, 1–12.

Buttsworth, D. R. 1996. Interaction of oblique shock waves and planar mixing regions. *J. Fluid Mech.* **306**, 43–57.

Buttsworth, D. R., Morgan, R. G. & Jones, T. V. 1995. A gun tunnel investigation of hypersonic free shear layers in a planar duct. *J. Fluid Mech.* **299**, 133–152.

Cabannes, H. 1960. Théorie des ondes de choc. *Encyclopedia of Physics* **9**(3), 162–224.

Cambon, C., Coleman, G. N. & Mansour, N. N. 1993. Rapid distortion analysis and direct simulation of compressible homogeneous turbulence at finite Mach number. *J. Fluid Mech.* **257**, 641–665.

Cassel, K. W., Ruban, A. I. & Walker, J. D. A. 1995. An instability in supersonic boundary-layer flow over a compression ramp. *J. Fluid Mech.* **300**, 265–285.

Cassel, K. W., Ruban, A. I. & Walker, J. D. A. 1996. The influence of wall cooling on hypersonic boundary-layer separation and stability. *J. Fluid Mech.* **321**, 189–216.

Caughey, D. A. 1982. The computation of transonic potential flows. *Annu. Rev. Fluid Mech.* **14**, 261–283.

Chang, C.-C. & Lei, S.-Y. 1996. An analysis of aerodynamic forces on a delta wing. *J. Fluid Mech.* **316**, 173–196.

Chang, C.-L. & Malik, M. R. 1994. Oblique-mode breakdown and secondary instability in supersonic boundary layers. *J. Fluid Mech.* **273**, 323–360.

Chaplygin, S. A. 1904. On gas jets. *Scientific Annals of the Imperial University of Moscow, Physico-mathematical division* **21**, 1–121 (in Russian). Translation in *NACA TM* 1063, 1944.

Cheng, H. K. 1993. Perspectives on hypersonic viscous flow research. *Annu. Rev. Fluid Mech.* **25**, 455–484.

Cherry, T. M. 1950. Exact solutions for flow of a perfect gas in a two-dimensional Laval nozzle. *Proc. Roy. Soc. London.* A **203**, 551–571.

Chisnell, R. F. 1998. An analytic description of converging shock waves. *J. Fluid Mech.* **354**, 357–375.

Chpoun, A., Passerel, D., Li, H. & Ben-Dor, G. 1995. Reconsideration of oblique shock wave reflections in steady flows. Part I. Experimental investigation. *J. Fluid Mech.* **301**, 19–35.

Clemens, N. T. & Mungal, M. G. 1995. Large-scale structure and entrainment in the supersonic mixing layer. *J. Fluid Mech.* **284**, 171–216.

Cole, J. D. & Cook, L. P. 1986. *Transonic Aerodynamics*. North-Holland.

Colella, P. & Henderson, L. F. 1990. The von Neumann paradox for the diffraction of weak shock waves. *J. Fluid Mech.* **213**, 71–94.

Coleman, G. N., Kim, J. & Moser, R. D. 1995. A numerical study of turbulent supersonic isothermal-wall channel flow. *J. Fluid Mech.* **305**, 159–183.

Colonius, T., Lele, S. K. & Moin, P. 1991. The free compressible viscous vortex. *J. Fluid Mech.* **230**, 45–73.

Colonius, T., Lele, S. K. & Moin, P. 1997. Sound generation in a mixing layer. *J. Fluid Mech.* **330**, 375–409.

Cope, W. F. 1953. Flow past bodies of revolution. *Modern Developments* **2**, 688–756.

Courant, R. & Friedrichs, K. O. 1948. *Supersonic Flow and Shock Waves.* Pure and Applied Mathematics, volume 1. Interscience.

Courant, R. & Hilbert, D. 1962. *Methods of Mathematical Physics. Volume 2. Partial Differential Equations.* Wiley.

Cowley, S. J. & Hall, P. 1990. On the instability of hypersonic flow past a wedge. *J. Fluid Mech.* **214**, 17–42.

Cox, E. A. & Kluwick, A. 1996. Resonant gas oscillations exhibiting mixed nonlinearity. *J. Fluid Mech.* **318**, 251–271.

Cramer, M. S. & Crickenberger, A. B. 1991. The dissipative structure of shock waves in dense gases. *J. Fluid Mech.* **223**, 325–355.

Cramer, M. S. & Sen., R. 1990. Mixed nonlinearity and double shocks in superfluid helium. *J. Fluid Mech.* **221**, 233–261.

Cramer, M. S. & Tarkenton, G. M. 1992. Transonic flows of Bethe–Zel'dovich–Thompson fluids. *J. Fluid Mech.* **240**, 197–228.

Crocco, L. 1937. Eine neue Strömungsfunktion für die Erforschung der Gase mit Rotation. *Zeit. Angew. Math. Mech.* **17**, 1–7.

Crocco, L. 1958. One-dimensional treatment of steady gas dynamics. *Princeton Series* **3**, 64–349.

Crow, S. C. & Bergmeier, G. G. 1996. Active sonic boom control. *J. Fluid Mech.* **325**, 1–28.

Curle, S. N. & Davies, H. J. 1971. *Modern Fluid Dynamics. Volume 2.* van Nostrand Reinhold.

Curran, E. T., Heiser, W. H. & Pratt, D. T. 1996. Fluid phenomena in scramjet combustion systems. *Annu. Rev. Fluid Mech.* **28**, 323–360.

Delale, C. F. & Crighton, D. G. 1998. Prandtl–Meyer flows with homogeneous condensation. Part 1. Subcritical flows. *J. Fluid Mech.* **359**, 23–47.

Delale, C. F., Schnerr, G. H. & Zierep, J. 1995. Asymptotic solution of shock tube flows with homogeneous condensation. *J. Fluid Mech.* **287**, 93–118.

Dillmann, A. 1994. Linear potential theory of steady internal supersonic flow with quasi-cylindrical geometry. Part 1. Flow in ducts. *J. Fluid Mech.* **281**, 159–191.

Dillmann, A. 1995. Linear potential theory of steady internal supersonic flow with quasi-cylindrical geometry. Part 2. Free jet flow. *J. Fluid Mech.* **286**, 327–357.

Ding, Z. & Gracewski, S. M. 1996. The behaviour of a gas cavity impacted by a weak or strong shock wave. *J. Fluid Mech.* **309**, 183–209.

Donovan, J. F., Spina, E. F. & Smits, A. J. 1994. The structure of a supersonic turbulent boundary layer subjected to concave surface curvature. *J. Fluid Mech.* **259**, 1–24.

Duck, P. W. 1990a. The inviscid axisymmetric stability of the supersonic flow along a circular cylinder. *J. Fluid Mech.* **214**, 611–637.

Duck, P. W. 1990b. The response of a laminar boundary layer in supersonic flow to small-amplitude progressive waves. *J. Fluid Mech.* **219**, 423–448.

Duck, P. W., Erlebacher, G. & Hussaini, M. Y. 1994. On the linear stability of compressible plane Couette flow. *J. Fluid Mech.* **258**, 131–165.

Duck, P. W. & Hall, P. 1990. Non-axisymmetric viscous lower-branch modes in axisymmetric supersonic flows. *J. Fluid Mech.* **213**, 191–201.

Durand, W. F., ed. 1934. *Aerodynamic Theory, Div. H.* Springer-Verlag.

Durbin, P. A. & Zeman, O. 1992. Rapid distortion theory for homogeneous compressed turbulence with application to modelling. *J. Fluid Mech.* **242**, 349–370.

Earnshaw, S. 1860. On the mathematical theory of sound. *Phil. Trans. Roy. Soc. London* **150**, 133–148.

Erlebacher, G., Hussaini, M. Y., & Shu, C.-W. 1997. Interaction of a shock with a longitudinal vortex. *J. Fluid Mech.* **337**, 129–153.

Erlebacher, G., Hussaini, M. Y., Speziale, C. G. & Zang, T. A. 1992. Toward the large-eddy simulation of compressible turbulent flows. *J. Fluid Mech.* **238**, 155–185.

Evvard, J. C. 1957. Diffusers and nozzles. *Princeton Series* **7**, 586–657.

Fedorchenko, A. T. 1997. A model of unsteady subsonic flow with acoustics excluded. *J. Fluid Mech.* **334**, 135–155.

Fernando, E. M. & Smits, A. J. 1990. A supersonic turbulent boundary layer in an adverse pressure gradient. *J. Fluid Mech.* **211**, 285–307.

Ferrari, C. 1996. Recalling the Vth Volta congress: High speeds in aviation. *Annu. Rev. Fluid Mech.* **28**, 1–9.

Ferrari, C. & Tricomi, F. G. 1968. *Transonic Aerodynamics.* Academic.

Ferri, A. 1949. *Elements of Aerodynamics of Supersonic Flows.* MacMillan.

Ferri, A. 1955a. The method of characteristics. *Princeton Series* **6**, 583–669.

Ferri, A. 1955b. Supersonic flows with shock waves. *Princeton Series* **6**, 670–748.

Ffowcs Williams, J. E. 1977. Aeroacoustics. *Annu. Rev. Fluid Mech.* **9**, 447–468.

Fox, M. D. & Kurosaka, M. 1996. Supersonic cooling by shock–vortex interaction. *J. Fluid Mech.* **308**, 363–379.

Frankl, F. I. 1945. On the problems of Chaplygin for mixed sub- and supersonic flows. *Akad. Nauk. SSSR, Izvestiya, Seriya Math.* **9**, 121–143.

Frick, C. W. 1957. The experimental aerodynamics of wings at transonic and supersonic speeds. *Princeton Series* **7**, 794–832.

Friedlander, F. G. 1958. *Sound Pulses.* Cambridge University Press.

Friedrichs, K. O. 1955. Mathematical aspects of flow problems of hyperbolic type. *Princeton Series* **6**, 31–60.

Fu, Y. & Hall, P. 1993. Effects of Görtler vortices, wall cooling and gas dissociation on the Rayleigh instability in a hypersonic boundary layer. *J. Fluid Mech.* **247**, 503–525.

Fu, Y. & Hall, P. 1994. Crossflow effects on the growth rate of inviscid Görtler vortices in a hypersonic boundary layer. *J. Fluid Mech.* **276**, 343–367.

Gallis, M. A. & Harvey, J. K. 1996. Modelling of chemical reactions in hypersonic rarefied flow with the direct simulation Monte Carlo method. *J. Fluid Mech.* **312**, 149–172.

Garrick, I. E. 1957. Nonsteady wing characteristics. *Princeton Series* **7**, 658–793.

Germain, P. 1964. Écoulements transsoniques homogènes. *Progress in Aeronautical Sciences* **5**, 143–273. Pergamon.

Germain, P. & Bader, R. 1952. Sur quelques problèmes relatifs à l'équation de type mixte de Tricomi. *Publ. Off. Nat. Étud. Aero. No. 54*, pp. 1–57.

Ghosh Choudhuri, P. & Knight, D. D. 1996. Effects of compressibility, pitch rate, and Reynolds number on unsteady incipient leading-edge boundary layer separation over a pitching aerofoil. *J. Fluid Mech.* **308**, 195–217.

Glass, I. I. & Sislian, J. P. 1994. *Nonstationary Flows and Shock Waves.* The Oxford Engineering Science Series **39**. Oxford University Press.

Glauert, H. 1928. The effect of compressibility on the lift of aerofoils. *Proc. Roy. Soc. London* A **118**, 113–119.

Gnoffo, P. A. 1999. Planetary-entry gas dynamics. *Annu. Rev. Fluid Mech.* **31**, 459–494.

Goddard, F. E. 1961. Supersonic tunnels. *Princeton Series* **8**, 491–532.

Goldshtein, A., Vainshtein, P., Fichman, M. & Gutfinger, C. 1996. Resonance gas oscillations in closed tubes. *J. Fluid Mech.* **322**, 147–163.

Goldstein, M. E. 1976. *Aeroacoustics.* McGraw-Hill.

Goldstein, M. E. & Wundrow, D. W. 1990. Spatial evolution of nonlinear acoustic mode instabilities on hypersonic boundary layers. *J. Fluid Mech.* **219**, 585–607.

Griffith, W. C. 1981. Shock waves. *J. Fluid Mech.* **106**, 81–101.

Griffith, W. C., Yanta, W. J. & Ragsdale, W. C. 1994. Supercooling in hypersonic nitrogen wind tunnels. *J. Fluid Mech.* **269**, 283–299.

Grosch, C. E. & Jackson, T. L. 1991. Inviscid spatial stability of a three-dimensional compressible mixing layer. *J. Fluid Mech.* **231**, 35–50.

Grove, J. W. & Menikoff, R. 1990. Anomalous reflection of a shock wave at a fluid interface. *J. Fluid Mech.* **219**, 313–336.

Grubin, S. E. & Trigub, V. N. 1993a. The asymptotic theory of hypersonic boundary-layer stability. *J. Fluid Mech.* **246**, 361–380.

Grubin, S. E. & Trigub, V. N. 1993b. The long-wave limit in the asymptotic theory of hypersonic boundary-layer stability. *J. Fluid Mech.* **246**, 381–395.

Guderley, K. G. 1942. Rückkehrkanten in ebener kompressibler Potentialströmung. *Zeit. Angew. Math. Mech.* **22**, 121–126.

Guderley, G. 1953. On the presence of shocks in mixed subsonic–supersonic flow patterns. *Adv. Appl. Mech.* **3**, 145–184.

Guderley, K. G. 1962. *The Theory of Transonic Flow.* Pergamon.

Guha, A. 1992. Jump conditions across normal shock waves in pure vapour-droplet flows. *J. Fluid Mech.* **241**, 349–369.

Gülen, S. C., Thompson, P. A. & Cho, H.-J. 1994. An experimental study of reflected liquefaction shock waves with near-critical downstream states in a test fluid of large molar heat capacity. *J. Fluid Mech.* **277**, 163–196.

Guo, Y. P. 1990. Sound generation by a supersonic aerofoil cutting through a steady jet flow. *J. Fluid Mech.* **216**, 193–212.

Gutmark, E. J. & Grinstein, F. F. 1999. Flow control with noncircular jets. *Annu. Rev. Fluid Mech.* **31**, 239–272.

Gutmark, E. J., Schadow, K. C. & Yu, K. H. 1995. Mixing enhancement in supersonic free shear flows. *Annu. Rev. Fluid Mech.* **27**, 375–417.

Haddad, O. M. & Corke, T. C. 1998. Boundary layer receptivity to free-stream sound on parabolic bodies. *J. Fluid Mech.* **368**, 1–26.

Hammerton, P. W. & Crighton, D. G. 1993. Overturning of nonlinear acoustic waves. Part 2. Relaxing gas dynamics. *J. Fluid Mech.* **252**, 601–615.

Hammerton, P. W. & Kerschen, E. J. 1996. Boundary-layer receptivity for a parabolic leading edge. *J. Fluid Mech.* **310**, 243–267.

Hayes, W. D. 1958. The basic theory of gasdynamic discontinuities. *Princeton Series* **3**, 416–481.

Hayes, W. D. 1971. Sonic boom. *Annu. Rev. Fluid Mech.* **3**, 269–290.

Hayes, W. D. & Probstein, R. F. 1966. *Hypersonic Flow Theory.* 2nd ed. Academic.

He, J., Kazakia, J. Y. & Walker, J. D. A. 1995. An asymptotic two-layer model for supersonic turbulent boundary layers. *J. Fluid Mech.* **295**, 159–198.

Heaslet, M. A. & Lomax, H. 1955. Supersonic and transonic small perturbation theory. *Princeton Series* **6**, 122–344.

Heister, S. D., McDonough, J. M., Karagozian, A. R. & Jenkins, D. W. 1990. The compressible vortex pair. *J. Fluid Mech.* **220**, 339–354.

Henderson, L. F. 1964. On the confluence of three shock waves in a perfect gas. *Aeronautical Quarterly* **15**, 181–197.

Henderson, L. F. 1965. The three-shock confluence on a simple wedge intake. *Aeronautical Quarterly* **16**, 42–54.

Henderson, L. F. 1987. Regions and boundaries for diffracting shock wave systems. *Zeit. Angew. Math. Mech.* **67**, 73–86.

Henderson, L. F., Colella, P. & Puckett, E. G. 1991. On the refraction of shock waves at a slow-fast gas interface. *J. Fluid Mech.* **224**, 1–27.

Henderson, L. F., Crutchfield, W. Y. & Virgona, R. J. 1997. The effects of thermal conductivity and viscosity of argon on shock waves diffracting over rigid ramps. *J. Fluid Mech.* **331**, 1–36.

Henderson, L. F. & Lozzi, A. 1975. Experiments on transition of Mach reflexion. *J. Fluid Mech.* **68**, 139–155.

Henderson, L. F. & Lozzi, A. 1979. Further experiments on transition to Mach reflexion. *J. Fluid Mech.* **94**, 541–559.

Henderson, L. F. & Menikoff, R. 1998. Triple-shock entropy theorem and its consequences. *J. Fluid Mech.* **366**, 179–210.

Holmes, R. L., Grove, J. W., & Sharp, D. H. 1995. Numerical investigation of Richtmyer–Meshkov instability using front tracking. *J. Fluid Mech.* **301**, 51–64.

Hornung, H. 1986. Regular and Mach reflection of shock waves. *Annu. Rev. Fluid Mech.* **18**, 33–58.

Hornung, H. G. 1998. Gradients at a curved shock in reacting flow. *Shock Waves* **8**, 11–21.

Howarth, L., ed. 1953. *Modern Developments in Fluid Dynamics: High Speed Flow. Volumes 1, 2.* Oxford University Press.

Howe, M. S. 1997. Edge, cavity and aperture tones at very low Mach numbers. *J. Fluid Mech.* **330**, 61–84.

Huang, P. G., Coleman, G. N. & Bradshaw, P. 1995. Compressible turbulent channel flows: DNS results and modelling. *J. Fluid Mech.* **305**, 185–218.

Hugoniot, H. 1889. Sur la propagation du mouvement dans les corps et spécialement dans les gaz parfaits. *Journal de l'École Polytechnique* **58**, 1–125.

Hyun, J. M. & Park, J. S. 1992. Spin-up from rest of a compressible fluid in a rapidly rotating cylinder. *J. Fluid Mech.* **237**, 413–434.

Igra, O., Falcovitz, J., Reichenbach, H. & Heilig, W. 1996. Experimental and numerical study of the interaction between a planar shock wave and a square cavity. *J. Fluid Mech.* **313**, 105–130.

Illingworth, C. R. 1953. Shock waves. *Modern Developments* **1**, 105–145.

Im, H. G., Bechtold, J. K. & Law, C. K. 1993. Analysis of thermal ignition in supersonic flat-plate boundary layers. *J. Fluid Mech.* **249**, 99–120.

Im, H. G., Helenbrook, B. T., Lee, S. R. & Law, C. K. 1996. Ignition in the supersonic hydrogen/air mixing layer with reduced reaction mechanisms. *J. Fluid Mech.* **322**, 275–296.

Ishii, R., Hatta, N., Umeda, Y. & Yuhi, M. 1990. Supersonic gas–particle two-phase flow around a sphere. *J. Fluid Mech.* **221**, 453–483.

Ivanov, M. S. & Gimelshein, S. F. 1998. Computational hypersonic rarefied flows. *Annu. Rev. Fluid Mech.* **30**, 469–505.

Jackson, T. L. & Grosch, C. E. 1990. Inviscid spatial stability of a compressible mixing layer. Part 2. The flame sheet model. *J. Fluid Mech.* **217**, 391–420.

Jackson, T. L. & Grosch, C. E. 1991. Inviscid spatial stability of a compressible mixing layer. Part 3. Effect of thermodynamics. *J. Fluid Mech.* **224**, 159–175.

Jacobs, J. W. 1992. Shock-induced mixing of a light-gas cylinder. *J. Fluid Mech.* **234**, 626–649.

Jacobs, J. W., Jenkins, D. G., Klein, D. L. & Benjamin, R. F. 1995. Nonlinear growth of the shock-accelerated instability of a thin fluid layer. *J. Fluid Mech.* **295**, 23–42.

Jameson, A. 1974. Iterative solution of transonic flows over airfoils and wings, including flows at Mach 1. *Comm. Pure Appl. Math.* **27**, 283–309.

Janzen, O. 1913. Beitrag zu einer Theorie der stationären Strömung kompressibler Flüssigkeiten. *Phys. Zeits.* **14**, 639–643.

Johnson, J. N. & Chéret, R. 1998. *Classic Papers in Shock Compression Science.* In the series High-Pressure Shock Compression of Condensed Matter (ed. R. A. Graham). Springer-Verlag.

Jones, R. T. & Cohen, D. 1957. Aerodynamics of wings at high speeds. *Princeton Series* **7**, 3–243.

Joseph, D. D., Huang, A. & Candler, G. V. 1996. Vaporization of a liquid drop suddenly exposed to a high-speed airstream. *J. Fluid Mech.* **318**, 223–236.

Jourdan, G., Houas, L., Haas, J.-F. & Ben-Dor, G. 1997. Thickness and volume measurements of a Richtmyer–Meshkov instability-induced mixing zone in a square shock tube. *J. Fluid Mech.* **349**, 67–94.

Kalkhoran, I. M., Smart, M. K. & Wang, F. Y. 1998. Supersonic vortex breakdown during vortex/cylinder interaction. *J. Fluid Mech.* **369**, 351–380.

Kantrowitz, A. R. 1958. One-dimensional treatment of nonsteady gas dynamics. *Princeton Series* **3**, 350–415.

Kerimbekov, R. M., Ruban, A. I. & Walker, J. D. A. 1994. Hypersonic boundary-layer separation on a cold wall. *J. Fluid Mech.* **274**, 163–195.

Kevlahan, N. K.-R. 1996. The propagation of weak shocks in non-uniform flows. *J. Fluid Mech.* **327**, 161–197.

Kevlahan, N. K.-R. 1997. The vorticity jump across a shock in a non-uniform flow. *J. Fluid Mech.* **341**, 371–384.

Khorrami, A. F. & Smith, F. T. 1994. Hypersonic aerodynamics on thin bodies with interaction and upstream influence. *J. Fluid Mech.* **277**, 85–108.

Klages, D. & Demmig, F. 1991. Model computations of the influence of carbon impurities on the ionization relaxation in krypton shock waves. *J. Fluid Mech.* **232**, 455–467.

Kluwick, A. 1993. Transonic nozzle flow of dense gases. *J. Fluid Mech.* **247**, 661–688.

Kluwick, A. & Scheichl, St. 1996. Unsteady transonic nozzle flow of dense gases. *J. Fluid Mech.* **310**, 113–137.

Konrad, W. & Smits, A. J. 1998. Turbulence measurements in a three-dimensional boundary layer in supersonic flow. *J. Fluid Mech.* **372**, 1–23.

Korobkin, A. 1992. Blunt-body impact on a compressible liquid surface. *J. Fluid Mech.* **244**, 437–453.

Korobkin, A. 1994. Blunt-body impact on the free surface of a compressible liquid. *J. Fluid Mech.* **263**, 319–342.

Korobkin, A. 1996a. Global characteristics of jet impact. *J. Fluid Mech.* **307**, 63–84.

Korobkin, A. 1996b. Acoustic approximation in the slamming problem. *J. Fluid Mech.* **318**, 165–188.

Kosinov, A. D., Maslov, A. A. & Shevelkov, S. G. 1990. Experiments on the stability of supersonic laminar boundary layers. *J. Fluid Mech.* **219**, 621–633.

Kuo, C.-Y. & Dowling, A. P. 1996. Oscillations of a moderately underexpanded choked jet impinging upon a flat plate. *J. Fluid Mech.* **315**, 267–291.

Kuo, Y. H. & Sears, W. R. 1955. Plane subsonic and transonic potential flows. *Princeton Series* **6**, 490–582.

References

Küssner, H. G. 1941. General airfoil theory. *NACA TM 979*.

Kutateladze, S. S., Nakoryakov, V. E. & Borisov, A. A. 1987. Rarefaction waves in liquid and gas–liquid media. *Annu. Rev. Fluid Mech.* **19**, 577–600.

Kynch, G. J. 1953. Blast waves. *Modern Developments* **1**, 146–157.

Landahl, M. T. 1961. *Unsteady Transonic Flow*. Pergamon.

Landau, L. D. 1944. Stability of tangential discontinuities in compressible fluid. *Akad. Nauk. SSSR, Comptes Rendus (Dokl.)* **44**, 139–141.

Landau, L. D. & Lifshitz, E. M. 1980. *Statistical Physics. Part I.* 3rd ed. Pergamon.

Landau, L. D. & Lifshitz, E. M. 1987. *Fluid Mechanics.* 2nd ed. Pergamon.

Laney, C. B. 1998. *Computational Gasdynamics*. Cambridge University Press.

Lee, S., Lele, S. K. & Moin, P. 1993. Direct numerical simulation of isotropic turbulence interacting with a weak shock wave. *J. Fluid Mech.* **251**, 533–562.

Lee, S., Lele, S. K. & Moin, P. 1997. Interaction of isotropic turbulence with shock waves: Effect of shock strength. *J. Fluid Mech.* **340**, 225–247.

Leib, S. J. 1991. Nonlinear evolution of subsonic and supersonic disturbances on a compressible free shear layer. *J. Fluid Mech.* **224**, 551–578.

Leib, S. J. & Lee, S. S. 1995. Nonlinear evolution of a pair of oblique instability waves in a supersonic boundary layer. *J. Fluid Mech.* **282**, 339–371.

Lele, S. K. 1994. Compressibility effects on turbulence. *Annu. Rev. Fluid Mech.* **26**, 211–254.

Leppington, F. G. & Sisson, R. A. 1997. On the interaction of a moving hollow vortex with an aerofoil, with application to sound generation. *J. Fluid Mech.* **345**, 203–226.

Lesser, M. B. & Field, J. E. 1983. The impact of compressible liquids. *Annu. Rev. Fluid Mech.* **15**, 97–122.

Levy, A., Ben-Dor, G. & Sorek, S. 1996. Numerical investigation of the propagation of shock waves in rigid porous materials: Development of the computer code and comparison with experimental results. *J. Fluid Mech.* **324**, 163–179.

Li, H. & Ben-Dor, G. 1997a. Analytical investigation of two-dimensional unsteady shock-on-shock interaction. *J. Fluid Mech.* **340**, 101–128.

Li, H. & Ben-Dor, G. 1997b. A parametric study of Mach reflection in steady flows. *J. Fluid Mech.* **341**, 101–125.

Liepmann, H. W. & Puckett, A. E. 1947. *Introduction to Aerodynamics of a Compressible Fluid.* Galcit Aeronautical Series. Wiley.

Liepmann, H. W. & Roshko, A. 1957. *Elements of Gasdynamics.* Wiley.

Lighthill, M. J. 1944. Two-dimensional supersonic aerofoil theory. *R & M 1929, Aerodynamics Department, National Physical Laboratory.* Also in *Collected Papers* **1** (1997).

Lighthill, M. J. 1947. The hodograph transformation in trans-sonic flow. Parts I–IV. *Proc. Roy. Soc. London* A **191**, 323–369, and **192**, 135–142 (with D. F. Ferguson). Also in *Collected Papers* **1** (1997).

Lighthill, M. J. 1949. Methods for predicting phenomena in the high-speed flow of gases. *J. Aeronaut. Sci.* **16**, 69–83. Also in *Collected Papers* **1** (1997).

Lighthill, M. J. 1953a. The hodograph transformation. *Modern Developments* **1**, 222–266.

Lighthill, M. J. 1953b. On boundary layers and upstream influence. I. A comparison between subsonic and supersonic flows. II. Supersonic flows without separation. *Proc. Roy. Soc. London* A **217**, 344–357 and 478–507. Also in *Collected Papers* **2** (1997).

Lighthill, M. J. 1955. Higher Approximations. *Princeton Series* **6**, 345–489. Also in *Collected Papers* **1** (1997).

Lighthill, M. J. 1956. Viscosity effects in sound waves of finite amplitude. In *Surveys in Mechanics* (ed. G. K. Batchelor and R. M. Davies), pp. 250–351. Cambridge University Press. Also in *Collected Papers* 1 (1997).

Lighthill, M. J. 1963. Introduction. Real and ideal fluids. Chapter 1 of *Laminar Boundary Layers* (ed. L. Rosenhead). Oxford University Press. Also in *Collected Papers* 2 (1997).

Lighthill, J. 1997. *Collected Papers of Sir James Lighthill*. Volumes 1–4 (ed. M. Y. Hussaini). Oxford University Press.

Likhterov, L. 1996. High-frequency acoustic noise emission excited by laser-driven cavitation. *J. Fluid Mech.* **318**, 77–84.

Liu, J. J. 1996. Sound wave structures downstream of pseudo-steady weak and strong Mach reflections. *J. Fluid Mech.* **324**, 309–332.

Lu, G. & Lele, S. K. 1993. Inviscid instability of a skewed compressible mixing layer. *J. Fluid Mech.* **249**, 441–463.

MacCormack, R. W. & Lomax, H. 1979. Numerical solution of compressible viscous flows. *Annu. Rev. Fluid Mech.* **11**, 289–316.

MacCrossan, M. N. 1990. Hypervelocity flow of dissociating nitrogen downstream of a blunt nose. *J. Fluid Mech.* **217**, 167–202.

MacCrossan, M. N. & Pullin, D. I. 1994. A computational investigation of inviscid hypervelocity flow of a dissociating gas past a cone at incidence. *J. Fluid Mech.* **266**, 69–92.

Mach, E. 1878. Über den Verlauf von Funkenwellen in der Ebene und im Raume. *Sitzungsber. Akad. Wiss. Wien* **78**, 819–838.

Mach, E. & Salcher, P. 1887. Photographische Fixierung der durch Projektile in der Luft eingeleiteten Vorgänge. *Sitzungsber. Akad. Wiss. Wien* **95**, 764–780.

Magi, E. C. & Gai, S. L. 1998. Flow behind castellated blunt-trailing-edge aerofoils at supersonic speeds. *J. Fluid Mech.* **375**, 85–111.

Mahesh, K., Lee, S., Lele, S. K. & Moin, P. 1995. The interaction of an isotropic field of acoustic waves with a shock wave. *J. Fluid Mech.* **300**, 383–407.

Mahesh, K., Lele, S. K. & Moin, P. 1997. The influence of entropy fluctuations on the interaction of turbulence with a shock wave. *J. Fluid Mech.* **334**, 353–379.

Mair, W. A. & Beavan, J. A. 1953. Flow past aerofoils and cylinders. *Modern Developments* 2, 612–687.

Majumdar, S. J. & Peake, N. 1998. Noise generation by the interaction between ingested turbulence and a rotating fan. *J. Fluid Mech.* **359**, 181–216.

Malik, M. R. & Spall, R. E. 1991. On the stability of compressible flow past axisymmetric bodies. *J. Fluid Mech.* **228**, 443–463.

Mallinson, S. G., Gai, S. L. & Mudford, N. R. 1997. The interaction of a shock wave with a laminar boundary layer at a compression corner in high-enthalpy flows including real gas effects. *J. Fluid Mech.* **342**, 1–35.

Manwell, A. R. 1971. *The Hodograph Equations. An Introduction to the Mathematical Theory of Plane Transonic Flow*. Oliver & Boyd.

Mashayek, F. 1998. Droplet–turbulence interactions in low-Mach-number homogeneous shear two-phase flows. *J. Fluid Mech.* **367**, 163–203.

McIntyre, T. J., Houwing, A. F. P., Sandeman, R. J. & Bachor, H.-A. 1991. Relaxation behind shock waves in ionizing neon. *J. Fluid Mech.* **227**, 617–640.

Menikoff, R. & Plohr, B. J. 1989. The Riemann problem for fluid flow of real materials *Reviews of Modern Physics* **61**, 75–130.

Messiter, A. F. & Matarrese, M. D. 1990. Hypersonic viscous interaction with strong blowing. *J. Fluid Mech.* **219**, 291–311.

Meyer, R. E. 1953. The method of characteristics. *Modern Developments* 1, 71–104.

Meyer, R. E. 1960. Theory of characteristics of inviscid gas dynamics. *Encyclopedia of Physics* **9** (3), 225–282.

Meyer, Th. 1908. Über zweidimensionale Bewegungsvorgänge in einem Gas, das mit Überschallgeschwindigkeit strömt. Dissertation, Göttingen. See also *Forschungsheft des Vereins deutscher Ingenieure* **62**, 31–67 (1908).

Mikhailov, V. V., Neiland, V. Ya. & Sychev, V. V. 1971. The theory of viscous hypersonic flow. *Annu. Rev. Fluid Mech.* **3**, 371–396.

Miles, J. W. 1959. *The Potential Theory of Unsteady Supersonic Flow.* Cambridge University Press.

Miller, M. F., Bowman, C. T. & Mungal, M. G. 1998. An experimental investigation of the effects of compressibility on a turbulent reacting mixing layer. *J. Fluid Mech.* **356**, 25–64.

Milne-Thomson, L. M. 1968. *Theoretical Hydrodynamics.* 5th ed. Macmillan.

Mitchell, B. E., Lele, S. K. & Moin, P. 1995. Direct computation of the sound from a compressible co-rotating vortex pair. *J. Fluid Mech.* **285**, 181–202.

Molenbroek, P. 1890. Über einige Bewegungen eines Gases bei Annahme eines Geschwindigkeitspotentials. *Archiv der Mathematik und Physik* **9**, 157–195.

Monaco, J. F., Cramer, M. S. & Watson, L. T. 1997. Supersonic flows of dense gases in cascade configurations. *J. Fluid Mech.* **330**, 31–59.

Mond, M. & Rutkevich, I. M. 1994. Spontaneous acoustic emission from strong ionizing shocks. *J. Fluid Mech.* **275**, 121–146.

Moody, D. M. 1990. Unsteady expansion of an ideal gas into a vacuum. *J. Fluid Mech.* **214**, 455–468.

Moore, D. W. & Pullin, D. I. 1998. On steady compressible flows with compact vorticity; the compressible Hill's spherical vortex. *J. Fluid Mech.* **374**, 285–303.

Moore, F. K. 1964. Hypersonic boundary layer theory. *Princeton Series* **4**, 439–527.

Morawetz, C. S. 1956, 1957. On the non-existence of continuous transonic flows past profiles. Parts 1 and 2. *Comm. Pure Appl. Math.* **9**, 45–68 and **10**, 107–131.

Moretti, G. 1987. Computation of flows with shocks. *Annu. Rev. Fluid Mech.* **19**, 313–337.

Morris, P. J. 1990. Instability waves in twin supersonic jets. *J. Fluid Mech.* **220**, 293–307.

Murman, E. M. & Cole, J. D. 1971. Calculation of plane steady transonic flows. *A.I.A.A.J.* **9**, 114–121.

Myers, M. R. & Kerschen, E. J. 1995. Influence of incidence angle on sound generation by airfoils interacting with high-frequency gusts. *J. Fluid Mech.* **292**, 271–304.

Myers, M. R. & Kerschen, E. J. 1997. Influence of camber on sound generation by airfoils interacting with high-frequency gusts. *J. Fluid Mech.* **353**, 221–259.

Naughton, J. W., Cattafesta, L. N. & Settles, G. S. 1997. An experimental study of compressible turbulent mixing enhancement in swirling jets. *J. Fluid Mech.* **330**, 271–305.

Nieuwland, G. Y. & Spee, B. M. 1973. Transonic airfoils: Recent developments in theory, experiment, and design. *Annu. Rev. Fluid Mech.* **5**, 119–150.

Ockendon, H. & Tayler, A. B. 1983. *Inviscid Fluid Flows.* Applied Mathematical Sciences **43**. Springer-Verlag.

Olejniczak, J., Wright, M. J. & Candler, G. V. 1997. Numerical study of inviscid shock interactions on double-wedge geometries. *J. Fluid Mech.* **352**, 1–25.

Oswatitsch, K. 1942. Kondensationserscheinungen in Überschalldüsen. *Zeit. Angew. Math. Mech.* **22**, 1–14.

Oswatitsch, K. 1956. *Gas Dynamics.* Applied Mathematics and Mechanics **1**. Academic.

Oswatitsch, K. 1959. Physikalische Grundlagen der Strömungslehre. *Encyclopedia of Physics* **8**(1), 1–124.

Pai, S.-I. 1959. *Introduction to the Theory of Compressible Flow.* Van Nostrand.

Panaras, A. G. 1992. Numerical investigation of the high-speed conical flow past a sharp fin. *J. Fluid Mech.* **236**, 607–633.

Panaras, A. G. 1997. The effect of the structure of swept-shock-wave/turbulent-boundary-layer interactions on turbulence modelling. *J. Fluid Mech.* **338**, 203–230.

Panda, J. 1998. Shock oscillation in underexpanded screeching jets. *J. Fluid Mech.* **363**, 173–198.

Parry, A. B. 1995. The effect of blade sweep on the reduction and enhancement of supersonic propeller noise. *J. Fluid Mech.* **293**, 181–206.

Paterson, A. R. 1983. *A First Course in Fluid Dynamics.* Cambridge University Press.

Paull, A., Stalker, R. J. & Mee, D. J. 1995. Experiments on supersonic combustion ramjet propulsion in a wind tunnel. *J. Fluid Mech.* **296**, 159–183.

Peake, N. 1992a. The interaction between a high-frequency gust and a blade row. *J. Fluid Mech.* **241**, 261–289.

Peake, N. 1992b. Unsteady transonic flow past a quarter-plane. *J. Fluid Mech.* **244**, 377–404.

Peake, N. 1993. The interaction between a steady jet flow and a supersonic blade tip. *J. Fluid Mech.* **248**, 543–566.

Peake, N. 1994a. The viscous interaction between sound waves and the trailing edge of a supersonic splitter plate. *J. Fluid Mech.* **264**, 321–342.

Peake, N. 1994b. The unsteady lift on a swept blade tip. *J. Fluid Mech.* **271**, 87–101.

Peake, N. 1997. On the behaviour of a fluid-loaded cylindrical shell with mean flow. *J. Fluid Mech.* **338**, 387–410.

Peake, N. & Crighton, D. G. 1991. Lighthill quadrupole radiation in supersonic propeller acoustics. *J. Fluid Mech.* **223**, 363–382.

Peake, N. & Kerschen, E. J. 1997. Influence of mean loading on noise generated by the interaction of gusts with a flat-plate cascade: Upstream radiation. *J. Fluid Mech.* **347**, 315–346.

Pedelty, J. A. & Woodward, P. R. 1991. Numerical simulations of the nonlinear kink modes in linearly stable supersonic slip surfaces. *J. Fluid Mech.* **225**, 101–120.

Peters, M. C. A. M., Hirschberg, A., Reijnen, A. J. & Wijnands, A. P. J. 1993. Damping and reflection coefficient measurements for an open pipe at low Mach and low Helmholtz numbers. *J. Fluid Mech.* **256**, 499–534.

Pham-Van-Diep, G. C., Erwin, D. A. & Muntz, E. P. 1991. Testing continuum descriptions of low-Mach-number shock structures. *J. Fluid Mech.* **232**, 403–413.

Pippard, A. B. 1957. *The Elements of Classical Thermodynamics.* Cambridge University Press.

Polachek, H. & Seeger, R. J. 1958. Shock wave interactions. *Princeton Series* **3**, 482–525.

Prakash, J. R. & Rao, K. K. 1991. Steady compressible flow of cohesionless granular materials through a wedge-shaped bunker. *J. Fluid Mech.* **225**, 21–80.

Prandtl, L. 1936. General considerations on the flow of compressible fluids. *NACA TM 805.*

Prasad, P. 1993. *Propagation of a Curved Shock and Nonlinear Ray Theory.* Pitman Research Notes in Mathematics **292**. Longman.

Prasad, P. & Ravindram, R. 1985. *Partial Differential Equations.* Wiley.

Quirk, J. J. & Karni, S. 1996. On the dynamics of shock-bubble interaction. *J. Fluid Mech.* **318**, 129–163.

Raman, G. 1997a. Screech tones from rectangular jets with spanwise oblique shock-cell structures. *J. Fluid Mech.* **330**, 141–168.

Raman, G. 1997b. Cessation of screech in underexpanded jets. *J. Fluid Mech.* **336**, 69–90.

Raman, G. & Taghavi, R. 1996. Resonant interaction of a linear array of supersonic rectangular jets: An experimental study. *J. Fluid Mech.* **309**, 93–111.

Raman, G. & Taghavi, R. 1998. Coupling of twin rectangular supersonic jets. *J. Fluid Mech.* **354**, 123–146.

Rankine, W. J. M. 1870. On the thermodynamic theory of waves of finite longitudinal disturbance. *Phil. Trans. Roy. Soc. London* **160**, 277–288.

Rayleigh, Lord 1910. Aerial plane waves of finite amplitude. *Proc. Roy. Soc. London* A **84**, 247–284.

Rayleigh, Lord 1916. On the flow of compressible fluid past an obstacle. *Phil. Mag.* **32**, 1–6.

Reichenbach, H. 1983. Contributions of Ernst Mach to fluid mechanics. *Annu. Rev. Fluid Mech.* **15**, 1–28.

Reisman, G. E., Wang, Y.-C. & Brennen, C. E. 1998. Observations of shock waves in cloud cavitation. *J. Fluid Mech.* **355**, 255–283.

Ribner, H. S. 1996. Effects of jet flow on jet noise by an extension to the Lighthill model. *J. Fluid Mech.* **321**, 1–24.

Rich, J. W. & Treanor, C. E. 1970. Vibrational relaxation in gas-dynamic flows. *Annu. Rev. Fluid Mech.* **2**, 355–396.

Riemann, B. 1860. Über die Fortpflanzung ebener Luftwellen von endlicher Schwingungsweite. *Abhandl. Ges. Wiss. Göttingen, Math.-Physik* **8**, 43–65.

Riesco–Chueca, P. & Fernández de la Mora, J. 1990. Brownian motion far from equilibrium: A hypersonic approach. *J. Fluid Mech.* **214**, 639–663.

Roe, P. L. 1986. Characteristic-based schemes for the Euler equations. *Annu. Rev. Fluid Mech.* **18**, 337–365.

Rossini, F. D. 1955. Fundamentals of thermodynamics. *Princeton Series* **1**, 3–110.

Rott, N. 1985. Jakob Ackeret and the history of the Mach number. *Annu. Rev. Fluid Mech.* **17**, 1–9.

Rusak, Z. 1993. Transonic flow around the leading edge of a thin airfoil with a parabolic nose. *J. Fluid Mech.* **248**, 1–26.

Rusak, Z. & Wang, C.-W. 1997. Transonic flow of dense gases around an airfoil with a parabolic nose. *J. Fluid Mech.* **346**, 1–21.

Rusanov, V. V. 1976. A blunt body in a supersonic stream. *Annu. Rev. Fluid Mech.* **8**, 377–404.

Ryzhov, O. S. 1978. Viscous transonic flows. *Annu. Rev. Fluid Mech.* **10**, 65–92.

Ryzhov, O. S. & Terent'ev, E. D. 1997. A composite asymptotic model for the wave motion in a steady three-dimensional subsonic boundary layer. *J. Fluid Mech.* **337**, 103–128.

Samtaney, R. & Zabusky, N. J. 1994. Circulation deposition on shock-accelerated planar and curved density-stratified interfaces: Models and scaling laws. *J. Fluid Mech.* **269**, 45–78.

Sandham, N. D. & Reynolds, W. C. 1991. Three-dimensional simulations of large eddies in the compressible mixing layer. *J. Fluid Mech.* **224**, 133–158.

Sarkar, S. 1995. The stabilizing effect of compressibility in turbulent shear flow. *J. Fluid Mech.* **282**, 163–186.

Sarkar, S., Erlebacher, G., Hussaini, M. Y. & Kreiss, H. O. 1991. The analysis and modelling of dilatational terms in compressible turbulence. *J. Fluid Mech.* **227**, 473–493.

Sasoh, A. & Takayama, K. 1994. Characterization of disturbance propagation in weak shock-wave reflections. *J. Fluid Mech.* **277**, 331–345.

Sauer, R. 1947. *Introduction to Theoretical Gas Dynamics.* J.W. Edwards.

Saunders, O. A. 1953. One-dimensional flow. *Modern Developments* **1**, 190–221.

Schiffer, M. 1960. Analytical theory of subsonic and supersonic flows. *Encyclopedia of Physics* **9**(3), 1–161.

Sears, F. W. & Salinger, G. L. 1975. *Thermodynamics, Kinetic Theory, and Statistical Thermodynamics.* 3rd ed. Addison–Wesley.

Sears, W. R. 1955. Small perturbation theory. *Princeton Series* **6**, 61–121.

Seddougui, S. O. & Bassom, A. P. 1997. Instability of hypersonic flow over a cone. *J. Fluid Mech.* **345**, 383–411.

Selerowicz, W. C., Szumowski, A. P. & Meier, G. E. A. 1991. Self-excited compressible flow in a pipe-collar nozzle. *J. Fluid Mech.* **228**, 465–485.

Séror, S., Zeitoun, D. E., Brazier, J.-Ph. & Schall, E. 1997. Asymptotic defect boundary layer theory applied to thermochemical non-equilibrium hypersonic flows. *J. Fluid Mech.* **339**, 213–238.

Serrin, J. 1959. Mathematical principles of classical fluid mechanics. *Encyclopedia of Physics* **8**(1), 125–263.

Shajii, A. & Freidberg, J. P. 1996. Theory of low Mach number compressible flow in a channel. *J. Fluid Mech.* **313**, 131–145.

Shapiro, A. H. 1953, 1954. *The Dynamics and Thermodynamics of Compressible Fluid Flow.* Volumes 1, 2. Ronald.

Shukhman, I. G. 1991. Nonlinear evolution of spiral density waves generated by the instability of the shear layer in a rotating compressible fluid. *J. Fluid Mech.* **233**, 587–612.

Simeonides, G. & Haase, W. 1995. Experimental and computational investigations of hypersonic flow about compression ramps. *J. Fluid Mech.* **283**, 17–42.

Simone, A., Coleman, G. N. & Cambon, C. 1997. The effect of compressibility on turbulent shear flow: A rapid-distortion-theory and direct-numerical-simulation study. *J. Fluid Mech.* **330**, 307–338.

Skews, B. W. 1997. Aspect ratio effects in wind tunnel studies of shock wave reflection transition. *Shock Waves* **7**, 373–383.

Skews, B. W., Atkins, M. D. & Seitz, M. W. 1993. The impact of a shock wave on porous compressible foams. *J. Fluid Mech.* **253**, 245–265.

Skews, B. W. & Takayama, K. 1996. Flow through a perforated surface due to shock-wave impact. *J. Fluid Mech.* **314**, 27–52.

Smith, D. R. & Smits, A. J. 1997. The effects of successive distortions on a turbulent boundary layer in a supersonic flow. *J. Fluid Mech.* **351**, 253–288.

Smith, F. T. & Brown, S. N. 1990. The inviscid instability of a Blasius boundary layer at large values of the Mach number. *J. Fluid Mech.* **219**, 499–518.

Smith, F. T. & Khorrami, A. F. 1991. The interactive breakdown in supersonic ramp flow. *J. Fluid Mech.* **224**, 197–215.

Smith, J. E. 1961. Transonic wind tunnels. *Princeton Series* **8**, 460–490.

Sobieczky, H. & Seebass, A. R. 1984. Supercritical airfoil and wing design. *Annu. Rev. Fluid Mech.* **16**, 337–363.

Solonnikov, V. A. & Kazhikhov, A. V. 1981. Existence theorems for the equations of motion of a compressible viscous fluid. *Annu. Rev. Fluid Mech.* **13**, 79–95.

Spiegelman, M. 1993. Flow in deformable porous media. Part 2. Numerical analysis – the relationship between shock waves and solitary waves. *J. Fluid Mech.* **247**, 39–63.

Spina, E. F., Donovan, J. F. & Smits, A. J. 1991. On the structure of high-Reynolds-number supersonic turbulent boundary layers. *J. Fluid Mech.* **222**, 293–327.

244 *References*

Spina, E. F., Smits, A. J. & Robinson, S. K. 1994. The physics of supersonic turbulent boundary layers. *Annu. Rev. Fluid Mech.* **26**, 287–319.

Squire, H. B. 1953. Heat transfer. *Modern Developments* **2**, 757–853.

Stalker, R. J. & Mudford, N. R. 1992. Unsteady shock propagation in a steady flow nozzle expansion. *J. Fluid Mech.* **241**, 525–548.

Stever, H. G. 1958. Condensation phenomena in high speed flows. *Princeton Series* **3**, 526–573.

Stewartson, K. 1964. *The Theory of Laminar Boundary Layers in Compressible Fluids.* Oxford University Press.

Stott, J. A. K. & Duck, P. W. 1994. The stability of a trailing-line vortex in compressible flow. *J. Fluid Mech.* **269**, 323–351.

Strykowski, P. J., Krothapalli, A. & Jendoubi, S. 1996. The effect of counterflow on the development of compressible shear layers. *J. Fluid Mech.* **308**, 63–96.

Stuart, J. T. 1986. Keith Stewartson: His life and work. *Annu. Rev. Fluid Mech.* **18**, 1–14.

Sturtevant, B. & Kulkarny, V. A. 1976. The focusing of weak shock waves. *J. Fluid Mech.* **73**, 651–671.

Sullivan, D. A. 1981. Historical review of real–fluid isentropic flow models. *Transactions of the ASME. Journal of Fluids Engineering* **103**, 258–267.

Sutherland, W. 1893. The viscosity of gases and molecular forces. *Phil. Mag.* **36**, 507–531.

Tam, C. K. W. 1995. Supersonic jet noise. *Annu. Rev. Fluid Mech.* **27**, 17–43.

Tam, C. K. W. & Ahuja, K. K. 1990. Theoretical model of discrete tone generation by impinging jets. *J. Fluid Mech.* **214**, 67–87.

Tam, C. K. W. & Auriault, L. 1998a. Mean flow refraction effects on sound radiated from localized sources in a jet. *J. Fluid Mech.* **370**, 149–174.

Tam, C. K. W. & Auriault, L. 1998b. The wave modes in ducted swirling flows. *J. Fluid Mech.* **371**, 1–20.

Tam, C. K. W. & Hu, F. Q. 1991. Resonant instability of ducted free supersonic mixing layers induced by periodic Mach waves. *J. Fluid Mech.* **229**, 65–85.

Tam, C. K. W. & Thies, A. T. 1993. Instability of rectangular jets. *J. Fluid Mech.* **248**, 425–448.

Tanaka, M. 1993. Mach reflection of a large-amplitude solitary wave. *J. Fluid Mech.* **248**, 637–661.

Taylor, G. I. 1910. The conditions necessary for discontinuous motion in gases. *Proc. Roy. Soc. London* A **84**, 371–377.

Taylor, G. I. & Maccoll, J. W. 1933. The air pressure on a cone moving at high speeds. *Proc. Roy. Soc. London* A **139**, 278–297 and 298–311.

Teipel, I. 1964. Ergebnisse der theorie schallnaher Strömungen. *Progress in Aeronautical Sciences* **5**, 104–142.

Temple, G. 1953. Unsteady motion. *Modern Developments* **1**, 325–374.

Thompson, P. A. 1985. *Compressible-Fluid Dynamics.* Maple.

Tijdeman, H. & Seebass, R. 1980. Transonic flow past oscillating airfoils. *Annu. Rev. Fluid Mech.* **12**, 181–222.

Timman, R. 1960. Linearized theory of unsteady flow of a compressible fluid. *Encyclopedia of Physics* **9**(3), 283–310.

Tirsky, G. A. 1993. Up-to-date gasdynamic models of hypersonic aerodynamics and heat transfer with real gas properties. *Annu. Rev. Fluid Mech.* **25**, 151–181.

Tomasini, M., Dolez, N. & Léorat, J. 1996. Instability of a rotating shear layer in the transonic regime. *J. Fluid Mech.* **306**, 59–82.

Tsien, H. S. 1939. Two-dimensional subsonic flow of compressible fluids. *J. Aeronaut. Sci.* **6**, 399–407.

Tsien, H. S. 1958. The equations of gas dynamics. *Princeton Series* **3**, 3–63.

Turkel, E. 1999. Preconditioning techniques in computational fluid dynamics. *Annu. Rev. Fluid Mech.* **31**, 385–416.

van den Berg, H. R., ten Seldam, C. A. & van der Gulik, P. S. 1993. Compressible laminar flow in a capillary. *J. Fluid Mech.* **246**, 1–20.

Van Dyke, M. 1975. *Perturbation Methods in Fluid Mechanics* (annotated edition). Parabolic.

Van Dyke, M. 1982. *An Album of Fluid Motion.* Parabolic.

Varley, E. & Cumberbatch, E. 1965. Non-linear theory of wave-front propagation. *J. Institute of Mathematics and its Applications* **1**, 101–112.

Vickers, I. P. & Smith, F. T. 1994. Theory and computations for breakup of unsteady subsonic or supersonic separating flows. *J. Fluid Mech.* **268**, 147–173.

Vincenti, W. G. & Kruger, C. H. 1965. *Introduction to Physical Gas Dynamics.* Wiley.

Virk, D., Hussain, F. & Kerr, R. M. 1995. Compressible vortex reconnection. *J. Fluid Mech.* **304**, 47–86.

von Kármán, Th. 1941. Compressibility effects in aerodynamics. *J. Aeronaut. Sci.* **8**, 337–356.

von Kármán, Th. 1955. On the foundation of high speed aerodynamics. *Princeton Series* **6**, 3–30.

von Mises, R. 1958. *Mathematical Theory of Compressible Fluid Flow.* Applied Mathematics and Mechanics **3**. Academic.

von Neumann, J. 1943. Oblique reflection of shocks. *Explosives Research Report No. 12, Navy Dept., Bureau of Ordnance, Washington D.C.* Also *Collected Works* **6**, 238–299 (1963). Pergamon.

von Ringleb, F. 1940. Exakte Lösungen der Differentialgleichungen einer adiabatischen Gasströmung. *Zeit. Angew. Math. Mech.* **20**, 185–198.

Vreman, A. W., Sandham, N. D. & Luo, K. H. 1996. Compressible mixing layer growth rate and turbulence characteristics. *J. Fluid Mech.* **320**, 235–258.

Vuillon, J., Zeitoun, D. & Ben-Dor, G. 1995. Reconsideration of oblique shock wave reflections in steady flows. Part 2. Numerical investigation. *J. Fluid Mech.* **301**, 37–50.

Wang, M. & Kassoy, D. R. 1990. Dynamic compression and weak shock formation in an inert gas due to fast piston acceleration. *J. Fluid Mech.* **220**, 267–292.

Wang, M. & Kassoy, D. R. 1992. Transient acoustic processes in a low-Mach-number shear flow. *J. Fluid Mech.* **238**, 509–536.

Wang, M., Lele, S. K. & Moin, P. 1996. Sound radiation during local laminar breakdown in a low-Mach-number boundary layer. *J. Fluid Mech.* **319**, 197–218.

Ward, G. N. 1953. Approximate methods. *Modern Developments* **1**, 267–324.

Ward, G. N. 1955. *Linearized Theory of Steady High-Speed Flow.* Cambridge University Press.

Watanabe, M. & Prosperetti, A. 1994. Shock waves in dilute bubbly liquids. *J. Fluid Mech.* **274**, 349–381.

Wells, V. L. & Renaut, R. A. 1997. Computing aerodynamically generated noise. *Annu. Rev. Fluid Mech.* **29**, 161–199.

Whitham, G. B. 1974. *Linear and Nonlinear Waves.* Wiley.

Wu, J.-Z. & Wu, J.-M. 1993. Interactions between a solid surface and a viscous compressible flow field. *J. Fluid Mech.* **254**, 183–211.

Yang, J., Kubota, T. & Zukoski, E. E. 1994. A model for characterization of a vortex pair formed by shock passage over a light-gas inhomogeneity. *J. Fluid Mech.* **258**, 217–244.

Young, A. D. 1953. Boundary layers. *Modern Developments* **1**, 375–476.

Young, J. B. & Guha, A. 1991. Normal shock-wave structure in two-phase vapour-droplet flows. *J. Fluid Mech.* **228**, 243–274.

Yu, K. H., Trouvé, A. & Daily, J. W. 1991. Low-frequency pressure oscillations in a model ramjet combustor. *J. Fluid Mech.* **232**, 47–72.

Zabusky, N. J. & Zeng, S. M. 1998. Shock cavity implosion morphologies and vortical projectile generation in axisymmetric shock–spherical fast/slow bubble interactions. *J. Fluid Mech.* **362**, 327–346.

Zel'dovich, Ya. B. & Raizer, Yu. P. 1966, 1967. *Physics of Shock Waves and High-Temperature Hydrodynamic Phenomena. Volumes 1, 2.* Academic.

Zel'dovich, Ya. B. & Raizer, Yu. P. 1969. Shock waves and radiation. *Annu. Rev. Fluid Mech.* **1**, 385–412.

Index

Takayama, K. 204, 231, 243
Taktakishvili, M. I. 230
Tam, C. K. W. 224, 226, 227, 244
Tanaka, M. 227, 244
Taran, J. P. 224, 231
Tarkenton, G. M. 233
Tayler, A. B. 224, 240
Taylor, G. I. 115, 223, 244
Taylor, H. S. 229
Taylor–Maccoll similarity solution 133
Teipel, I. 224, 244
temperature
  absolute 23, 30
  gradient 20
  thermodynamic 30
Temple, G. 223, 244
Temple's solution 220
Terekhin, V. E. 230
Terent'ev, E. D. 242
thermal conduction 28
thermal expansion, coefficient of 37, 49, 53
thermodynamic change
  adiabatic 31
  irreversible 31, 32, 34
  isentropic 31, 36
  nonisentropic 36
  reversible 30–2, 34
thermodynamic derivative
  enthalpy 43
  Gibbs function 45
  Helmholtz function 44
  perfect gas 48
thermodynamic equilibrium 32
thermodynamic formulae
  arbitrary fluid 37–45
  perfect gas 45–50
  polytropic gas 50–3
  practical 39
  reference collection 39
  standard form 40
thermodynamic potential 33
thermodynamic temperature 30
thermodynamic variable 20–1, 29, 31, 33
thermodynamics
  first law 31
  second law 30–1
  third law 24, 31
Thies A. T. 244
Thompson, P. A. 54, 115, 224,
    235, 244
throat 123
Tijdeman, H. 160, 224, 244
Timman, R. 223, 244
Tirsky, G. A. 224, 244
Toepler, A. 3
Tomasini, M. 227, 244
total energy 58
total head 58

transition curve in hodograph method 215–19
transonic
  equation 157, 158
  similarity solutions 163
  flow past aerofoil 4, 153
  nozzle 131
  similarity parameter 114, 152, 159
  small-disturbance equation 160, 163
  transport equation 82–3
Treanor, C. E. 224, 242
Tricomi, F. G. 160, 224, 234
Tricomi's equation 209
Trigub, V. N. 235
triple point 129, 180
triple-shock entropy theorem 202
Trouvé, A. 246
Truesdell, C. A. 6, 229
Tsien, H. S. 64, 223, 244, 245
Tukey, J. W. 5
turbine 5
turbulence 5, 227
Turkel, E. 224, 245

Umeda, Y. 236
underexpanded jet 130, 140
United Technologies 5
upwind differencing 160

vacuum condition 122, 216
vacuum-condition circle 217
Vainshtein, P. 235
van den Berg, H. R. 227, 245
van der Gulik, P. S. 245
Van Dyke, M. 6, 115, 204, 224, 245
variable
  non-thermodynamic 21
  selected 21
  of state 20
  thermodynamic 20–1, 29, 31, 33
Varley, E. 83, 245
velocity gradient 19–21
vena contracta 125–6
vibrational energy 51
Vickers, I. P. 226, 245
Vincenti, W. G. 54, 224, 245
Virgona, R. J. 204, 236
Virk, D. 227, 245
viscosity 19, 28
  bulk 20
  nonnegative 23, 32
  shear 19
  temperature dependence 20
Volta Congress, Vth 5
volume, specific 21
von Kármán, Th. 223, 245
von Mises, R. 220, 221, 224, 245
von Neumann, J. 4, 115, 204, 223, 245
von Neumann

Index

Printed in the United States
By Bookmasters